HORNET PLUS THREE

THE STORY OF THE *APOLLO 11* RECOVERY

BOB FISH

Reno, Nevada

Creative Minds Press
an imprint of **Beagle Bay, Inc.**
Reno, Nevada
info@beaglebay.com

Copyright ©2009 Bob Fish

All rights reserved. No part of this book may be reproduced or transmitted in any form or by any means, electronic or mechanical, including photocopying, recording, or by any informational storage and retrieval system, without permission in writing from the publisher, except for the inclusion of brief quotations in a review.

Visit our websites at:
http://www.uss-hornet.org.
http://www.creativemindspress.com; http://www.beaglebay.com

Book Design: Robin P. Simonds
Editing: Jacqueline Church Simonds

Front cover photograph by Milt Putnam from a Navy helicopter during an exercise prior to the recovery of *Apollo 11*.
Hornet "Apollo Recovery Ship" artwork by James Farrell of West Coast Novelty.
Author photos by Carol Lee.

Library of Congress Cataloging-in-Publication Data

Fish, Bob.
 Hornet plus three : the story of the Apollo 11 recovery / Bob Fish.
   p. cm.
 Includes bibliographical references and index.
  ISBN 978-0-9749610-7-1 (hardcover : alk. paper) 1. Apollo 11 (Spacecraft) 2. Project Apollo (U.S.) 3. Space rescue operations. 4. Moon--Exploration. 5. Space flight to the moon. 6. Hornet (Aircraft carrier : CVS 12) I. Title.

TL789.8.U6F58 2009
629.45'4--dc22

                    2008053280

First Edition
Printed in China
14 13 12 11 10 09   1 2 3 4 5 6 7 8 9 10

Dedication

    Landing a man on the Moon, and returning him safely to Earth, is undoubtedly America's greatest peacetime technological achievement. Several hundred thousand people were directly involved in some aspect of this important national program, whether serving as a NASA employee, government contractor or member of the military.

    This book is dedicated to all military personnel who served during that era, whether actively fighting in the armed conflict in Vietnam or maintaining the constant global vigilance required during the Cold War. The success of this amazing lunar landing program changed the course of history and could not have been achieved without their contributions.

    More personally, this work is dedicated to my wife, Jennifer, for her understanding, patience and support for this eight-year project. She participated in many of my trips around the country to collect information, endured my absence for too many meetings aboard the Hornet Museum and understood the endless hours I was sequestered in front of my computer.

Semper Fi

Acknowledgments

A few individuals made significant personal contributions to the development of this book, not the least of which was inspiring me to work on it for eight years! Each has provided valuable insight, an eyewitness account, or hard-to-find background materials. Without their support, this book could not have accomplished its goal of documenting the numerous facets of the DoD support for America's lunar landing program nor the *Apollo 11* recovery operation. Listed in order of their contribution effort:

RADM Carl J. Seiberlich USN (Ret.)	Commanding Officer - USS *Hornet Apollo 11* Recovery Force Commander
Captain Charles B. Smiley USN (Ret.)	Executive Officer - HS-4
John Stonesifer	NASA (Branch Chief, Recovery Systems)
Commander Clancy Hatleberg USN (Ret.)	UDT Recovery Team & Decontamination Swimmer
PHCS (AC) Robert Lawson (Ret.)	Chief Photographer - USS *Hornet*
PHC (AC) Milt Putnam (Ret.)	Photojournalist - HS-4
Captain Chris Lamb USN (Ret.)	Executive Officer - USS *Hornet*
Neil Armstrong	NASA - Commander *Apollo 11*
VADM Donald S. Jones USN (Ret.)	Commanding Officer - HS-4
Don Blair	Mutual Radio Network - *Apollo 11* Broadcast Reporter
Rolf Sabye	Quartermaster - USS *Hornet*
John Blossom	MELPAR - MQF Program Manager
Tim Wilson	USS *Hornet* - Public Affairs Officer
Mike Wheat	USS *Hornet* - *Apollo 11* Recovery Cruise Book

Many other individuals contributed time, photos, recollections or other information that helped make this book possible. The author wishes to express sincere appreciation to each of them for their kindness and support. Special thanks go to Mike Gentry and the NASA Media Resource Center at Johnson Space Center for their high level of support throughout this project.

RADM Warren E. Aut USN (Ret.)	James Arnold	Robert Burns
Judy Allton	Eric Beheim	Robert Carlisle

Pete Clayton
Dwayne Day
Michael Esslinger
Frank Farmer
Alfred Hassebrock
RADM John Higginson USN (Ret.)
Charles Hinton
John Hirasaki
Kenny Hoback
Paul Hutchinson
Eric Jones
Alan Krohn

Jack LaBounty
Roger Launius
ADM Charles R. Larson USN (Ret.)
Carol Lee
Jim Lewis
Sy Liebergot
Colin Mackellar
Jim McDade
BGEN Daniel B. Mcdyre USMC (Ret.)
Christopher Miller
Howard Mooney
Captain Hugh Murphree USN (Ret.)

Allan Needell
Susan Phipps
Richard Powers
Paul Rambaut
Gloria Sanchez
Crystal Schroeder
Bob Shanahan
Thomas Sheehan
Glen Swanson
Jim White
Marty Wilson

Table of Contents

Acknowledgments v

Foreword xi

Preface xiii

Introduction xv

Chapter One
    Spacecraft Recovery Program 5
        Department of Defense Support 7
        DDMS Recovery Support Overview 8
        U.S. Air Force Overview 9
        U.S. Navy Overview 9

Chapter Two
    The Learning Curve 13
        Initial Preparations 13
        Project Mercury 15
        Project Gemini 19

Chapter Three
    DoD Apollo Mission Support 23
        Apollo Command Module Recovery Concept 23
        Final DoD Preparations 25
        Nominal DoD Mission Deployment 28
        Technical Note: MSFN Mobile Facilities 31

Chapter Four
Initial Apollo Fights ... 36
The First Lunar Mission ... 37
One Final Earth Test ... 40
The Second Lunar Mission ... 41

Chapter Five
Earth Contamination Issues ... 45
Mobile Quarantine Facility (MQF) ... 47
Biological Isolation Garment ... 50
Biological Decontamination Process ... 51
Quarantine Training ... 52
Technical Note: MQF R&D Program ... 54

Chapter Six
Primary Recovery Force: Ships ... 59
USS *Hornet* (CVS-12) ... 59
USS *Goldsborough* (DDG-20) ... 61
USS *Hassayampa* (AO-145) ... 64
USS *Arlington* (AGMR-2) ... 64

Chapter Seven
Primary Recovery Force: Units ... 67
Shipboard Units ... 67
Non-Shipboard Support Units ... 75
Technical Note: Navigational Aids for Locating the Command Module ... 79

Chapter Eight
Pre-Flight Preparations ... 81

Chapter Nine
Preparing for the President ... 90

Chapter Ten
Flight of *Apollo 11* ... 99

Chapter Eleven
    Splashdown and Recovery 108

Chapter Twelve
    Return to Pearl Harbor 134

Epilogue

Appendix A
    Personal Recollections 151
        Neil Armstrong 151
        John Stonesifer 153
        Captain Charles B. Smiley 156
        Clancy Hatleberg 159
        Don Blair 162
        Milt Putnam 165
        Rear Admiral Carl J. Seiberlich 168

Appendix B
    Time-lines 177
        Time-line of the "Space Race" 177
        Time-line of *Apollo 11* Recovery 179

Appendix C
    Key Speeches 183
        President John F. Kennedy's speech before Congress, May 25, 1961. 183
        President John F. Kennedy's speech at Rice University in Houston, Texas, on September 12, 1962.. 185
        President Richard M. Nixon's Telephone Conversation with the Astronauts on the Moon, July 20, 1969. 189
        President Richard M. Nixon's Prepared Speech, In the event of an *Apollo 11* Disaster (written by William Safire) . . 190
        President Richard M. Nixon Congratulates the Returned Astronauts Aboard the *Hornet*, July 24, 1969 191

Appendix D
    Key Crew Assignments 194

Appendix E
    Photography Related Information 196
Glossary of Acronyms & Abbreviations 202
Bibliography 205
About the Author 207
Index 209

Foreword

Young people today ask, "Why spacecraft recovery at sea?" when all they have known is the space shuttle returning to Earth and landing on a runway. The Mercury, Gemini and Apollo spacecraft were incapable of landing on land and were designed to be recovered by parachute and a water landing for a myriad of reasons: launch abort area, structural limitations, booster capabilities and mostly, because of all the water landing areas on Earth.

As a former naval aviator, I am intimately familiar with the normal routine of aircraft carrier operations and naval ship deployments. This gives me a keen insight into how unusual the recovery evolutions were for the primary recovery forces and all the training and preparation work that went into ensuring a successful recovery. After my *Gemini 11* flight in 1966, Pete Conrad and I were plucked from the Atlantic by very professional UDT and helicopter teams based on the USS *Guam*, a helicopter assault ship.

In thinking about writing this foreword, I researched my memory bank and remember that the crew of *Apollo 12* sent a message to Rear Admiral Donald C. Davis, Commander Task Force 130. I remembered writing the message in one of the onboard checklists. Having been found, it reads:

RADM Davis
Recovery Force
USS *Hornet*

Dear Red Dog,
Apollo 12, with three tail-hookers, expect recovery ship to make its PIM as we have energy for only one pass.
Pete, Dick and Al

As any carrier pilot would know, PIM stands for "Point of Intended Movement," and lets the pilot know where the ship will be for recovery.

It is assumed that Mission Control Center (MCC) Houston passed our message on to Rear Admiral Davis. USS *Hornet* made its PIM and *Apollo 12* splashed down within 2.5 miles of the ship, allowing personnel to view the entire recovery procedure.

During landing, *Apollo 12* experienced a significant sea-state and encountered a "hard landing," one that was unexpected and caused a minor injury to Al Bean—our only casualty.

Recovery operations were "by the book," as we expected, knowing the professionalism and dedication of our fellow recovery forces.

As the crewmembers of *Apollo 12* in 1969, Pete, Alan and I endured the lunar quarantine version of a mission recovery—well described in this book—which again, was conducted masterfully by the USS *Hornet* and its embarked recovery forces.

When we arrived at our MQF we saw a sign over the door that read, "Three More Like Before." This sign was emblematic that just like the recovery of *Apollo 11*—safety, discipline and professionalism were paramount, and once again men were returned safely to Earth.

Upon the ship's arrival at Pearl Harbor, and after the MQF was off-loaded, Admiral John McCain, CINCPAC, presented all three new Navy captains with their collar devices.

Hornet Plus Three shines a richly-deserved light on an area often overlooked by history books in regards to the terrific Moon race of the 1960's. Thousands of young military personnel, led by officers such as then-Captain Carl Seiberlich and dedicated NASA personnel, were a critical element in fulfilling President Kennedy's directive of "returning them safely to Earth." The fact that they were fighting a Cold War around the world, and a "hot war" in Viet Nam at the same time, makes their accomplishments all the more special.

Bob Fish promised Rear Admiral Seiberlich that he would tell the story of USS *Hornet* and her involvement in recovery operations during the Apollo Program. This he has done. As the first in-depth book written on "spacecraft recovery" it is likely to become a keystone of the legacy for all the Navy and NASA personnel who executed these missions, and I have thoroughly enjoyed reading it.

In the vernacular of the Naval Services:

Well done, Bob Fish.

Richard Gordon, Captain, USN (Ret.), *Gemini 11, Apollo 12*

Preface

In the course of modern world history, one attribute of a democratic society has become glaringly apparent. Significant achievements happen when three critical elements are present:
- A strategic vision of what needs to happen,
- A consensus roadmap of how to make it happen,
- A critical mass of ordinary citizens who *choose* to make it happen.

In early 1961, President Kennedy sensed the competition for global leadership between Democracy and Communism would not be won on the battlefield. The concept of "Mutually Assured Destruction" (MAD) prevented either Superpower from inflicting a total military defeat on the other without suffering a similar fate. During a speech to Congress that May, the President outlined a bold new vision for America that would showcase its scientific and technological prowess. Our country became committed to landing a man on the Moon, and returning him safely to the Earth, before the end of the decade. In an instant, a significant percentage of the Cold War battleground shifted into the field of space exploration, rather than development of new weapons of mass destruction.

The thoughts of all mankind were focused as never before when Neil Armstrong jumped from the lunar landing module to the surface of the Moon on July 20, 1969. Much of humanity shared a common experience of awe during that first lunar surface walk.

However, there was still a pesky final detail that had to be managed successfully—returning these first Moonwalkers safely to Earth. While there was a massive DoD and NASA planning and support effort involved, in the end this activity fell onto the shoulders of Captain Carl Seiberlich and the recovery team aboard the USS *Hornet*. The participants spanned a full generation in range of ages, most being born after World War II had ended, yet all were clearly an extension of the "greatest generation" of Americans. They worked together as a team, focused on a common objective and ensured that at the end of a very special day, they could say with pride "job well done."

This is their story.

Introduction

I grew up in Orlando, Florida during the 1960's; at that time it was a fairly sleepy Southern town. However, older Floridians of that era were acutely aware of the contest between Democracy and Communism for global dominance. Tensions were high as the Cold War escalated. My family had a front row seat when the Cuban Missile Crisis with the Soviet Union almost boiled over in 1962. One Fall day, our house shook violently for over an hour as waves of B-52 bombers and KC-135 tankers took off from McCoy Air Force Base at full military throttle and screamed overhead. My mother and father, both World War II veterans of the U.S. Marine Corps, were frightened by the thought of "Mutually Assured Destruction" that a nuclear war would bring. For me, the issue came down to having to wash the kerosene mist released from the jet engines off the family car after the scramble was over. It didn't really occur to me we had survived a potentially devastating, eyeball-to-eyeball political event.

On May 25, 1961 President John F. Kennedy spoke to a joint session of Congress about the urgent national need for the United States not to cede control of space to the Soviet Union. In the course of this speech, he made the now oft-repeated challenge, "I believe that this nation should commit itself to achieving the goal, before this decade is out, of landing a man on the Moon and returning him safely to the Earth."

As the 1960's wore on, the U.S. government found less stressful ways to interrupt our mundane lives by launching rockets from Cape Canaveral (now called Kennedy Space Center). The unmanned test or satellite missions occurred without much attention, unless they blew up on the launch pad. However, the manned missions were filled with drama since human lives were at stake. Each launch provided us a spectacular scene, carved from the ongoing global struggle with the Soviet Union, and took our minds off the growing war in Southeast Asia.

For manned flights of the lunar space program, Mercury, Gemini and Apollo, the countdown and launch were broadcast over television. If a launch occurred during the school day, our class would be interrupted for the final two minutes of the countdown. When the TV announcer declared, "We have liftoff," all the kids raced outside and faced east toward the launch complex. We watched in awe as the glistening sliver of metal arced into the sky on a huge tongue of flame and smoke.

The launch of the mighty Saturn V rockets of the Apollo program was accompanied by a minor shaking of the ground. It seemed the Earth itself was helping to heave the astronauts into space, cheering along mankind's greatest scientific endeavor. The idea of putting a man on the Moon was so fantastic and exciting that it united much of the world's population together with a common hope.

As the decade wore on, my attention gradually focused elsewhere. I went away to the University of Virginia to engage in all things "collegian"—playing Frisbee, dating, attending football games, learning how to drink responsibly, etc. Because of the military draft, "the elephant lurking in the room" for all young people in the late 1960s was the other epic struggle against Communism, the Vietnam War. After closing out my college adventure, I followed in my family's footsteps and joined the Marine Corps.

On July 20, 1969, in the barracks at MCB Quantico, Virginia, I watched a small black and white TV screen as it showed grainy images of Neil Armstrong and Buzz Aldrin walking on the Moon. The range of emotions I felt was very complex. It seemed surreal to think mankind was now capable of leaving our native home and journeying into the heavens. The images from all the Apollo missions made our planet seem smaller, our differences less important and our population more humble. I also had a great sense of pride because humans had shown what great achievements can be made if there is a clearly stated goal and public consensus for attaining it. However, the rest of the Moon landing program became a blur to me, while I finished my enlistment on Okinawa working six days a week in a secret USMC data center.

Like everyone else, I had witnessed something spectacular in the history of mankind but had to get on with an everyday life, largely ignorant of exactly how this lofty goal was accomplished.

Putting a man on the Moon and returning him safely to Earth was then, and remains now, a very difficult and complex process. NASA needed a host of new technologies, many not yet on the drawing board when President Kennedy gave his speech. They welded together an army of employees, contractors and other organizations and kept them focused by using advanced program management and organizational communication techniques. The NASA team performed brilliantly in accomplishing their engineering and management tasks. They also relied on a lot of help from the Department of Defense (DoD).

America's leadership in Washington D.C. wisely wanted to ensure the public focus of our space effort was civilian in nature, albeit running in parallel with military needs. To support their own research and development of advanced missile and supersonic aircraft programs in the late 1950's, the DoD established a worldwide array of tracking, relay, recovery, logistics and other support assets. For every NASA space flight, whether sub-orbital, Earth-orbital or lunar-orbital in nature, many of these facilities were placed into support roles. The NASA public relations machine worked intensively to maintain high levels of public interest, which lead to continued Congressional funding for the Moon race. What the world remembers now about the 1960's space exploration activities largely involves the civilian space effort while the DoD effort remained in the shadows.

One of the most important DoD support functions was the landing and recovery aspect of a space flight—the part about "returning him safely to the Earth."

As a result, the mainstream U.S. Navy stepped into the Moon race milieu, while fighting an active war in Vietnam and a Cold War everywhere else. The officers and crews of various ships were thrust into the spotlight of the world stage, if only for a week or two until more pressing news relegated them to history. At any point in time, a variety of ships could be re-tasked to support a NASA mission and any of their commanding officers could be called upon to ensure a safe return to Earth of the astronauts.

One such ship was the aircraft carrier USS *Hornet* (CVS-12), a veteran of World War II with nine battle stars and a Presidential Unit Citation. The latter was awarded to the ship and its crew as a result of fifteen months of continuous raging combat action in the Pacific.

And one such officer was Captain Carl J. Seiberlich, also a decorated World War II veteran. He graduated from the U.S. Merchant Marine Academy at King's Point, New York in 1943. During the war, he served in both the Atlantic and Pacific theaters and was the navigator onboard the destroyer USS *Mayo* in Tokyo Bay on the day of Japan's surrender. After the war, he decided to pursue a career in aviation and gained his first pair of aviator wings in "lighter-than-air" craft (i.e., a blimp). When he reported aboard *Hornet* as Commanding Officer on May 23, 1969 his Navy career reached its zenith.

The teak-decked *Hornet's* career was nearing its end, however. On May 2, 1969 she returned to her home port of Long Beach, California from a third tour of duty off the coast of Vietnam. To make way for more capable nuclear powered aircraft carriers, the Navy had scheduled her to be mothballed in June 1970. It appeared her final year's activities would be spent performing basic carrier qualifications for new aviators and conducting anti-submarine warfare research.

Then on June 1, the Navy announced that *Hornet* had been selected as the Primary Recovery Ship (PRS) for *Apollo 11* and Captain Seiberlich would be the Commander of all Primary Recovery Forces in the Pacific. The recovery date for this historic lunar landing mission was less than two months away, and there was little time to be wasted in preparing the ship and crew.

Most of what transpired in the following eight weeks has never been told publicly. Ship movements and personnel assignments in the Pacific were highly guarded secrets during the Vietnam War and the ongoing silent battle with Soviet missile submarines. After the Apollo program, NASA and the space industry moved on to the Shuttle program, while much of the Pacific Fleet documentation from the 1960s was boxed up and placed in secure storage.

In 1998, the former U.S. Navy aircraft carrier USS *Hornet* (CV-12) was dedicated as a public museum. The featured speaker for the grand opening was Buzz Aldrin, the second human to walk on the Moon. Having a former USAF test pilot and lunar astronaut as the keynote speaker certainly emphasized the unique heritage of *Hornet*, the Prime Recovery Ship for *Apollo 11* and *Apollo 12*.

The museum fully embraces this aspect of the ship's history. In 2000, when I joined the museum as Apollo Curator, the Manned Spacecraft Recovery Research Center was established with a specific goal of investigating, documenting and disseminating information about the Navy's support of NASA operations in the pre-Shuttle era. Over time, the museum built a one-of-a-kind exhibit, creating an *Apollo 11* recovery diorama in Hangar Bay #2. This exhibit contains all the major elements from July 24, 1969: a Mobile Quarantine Facility, a Sea King helicopter, a Biological Isolation Garment and a flown Apollo Command Module. Every year, the museum commemorates the recoveries of the two initial lunar landing missions.

In 2004, the museum made history by hosting a ten day event to mark the thirty-fifth anniversary of the flight of *Apollo 11*. For the first time since 1969, many of the key participants were brought together to discuss this seminal event with the media and the public. That same year, the museum assisted in the creation of the first major book devoted to spacecraft recovery operations entitled *Splashdown: NASA and the Navy*. It was written by Don Blair, the radio announcer who broadcast the *Apollo 11* recovery live from the ship in 1969. The book's inaugural release was held at the Hornet Museum amid great fanfare.

Hornet Plus Three: The Story of the Apollo 11 Recovery documents the overall DoD support infrastructure as well as providing an in-depth look at the *Apollo 11* recovery operation. This book is not just about the equipment used and timelines followed. It is intended to inspire future generations to become involved in the military or space exploration programs for science and technology purposes. Significant material concerning the participants themselves is included—insights that can only come from having had a leadership position in this momentous event.

Two of the men in these key roles have recently passed away. Rear Admiral Carl Seiberlich was the final Commanding Officer of the USS *Hornet* and the last man to leave the ship when she was decommissioned. Vice Admiral Don Jones, the Commanding Officer of ASW helicopter squadron HS-4, was the primary recovery pilot for *Apollo 8* and *Apollo 11*. Before they died, I had the opportunity to interview both men.

Admiral Seiberlich provided many details of the recovery activities with great clarity and insight. Those stories rekindled distant memories of mine about America's technological prowess and widespread public enthusiasm during the Moon race. I realized how much I really didn't know about the Apollo program, other than the fiery ascent during the launch.

In my conversations with these men and other members of the recovery team, every person echoed the same two sentiments. First, they were just "doing their duty," making sure the recovery operation went according to plan with no casualties or problems. Second, in spite of incredible military and civilian careers afterward, they all said it was a defining moment in their lives and that the Apollo recovery team (Navy, NASA, contractor, media, etc) was the finest group of individuals they had ever worked with.

Just days before he passed away at the age of eighty-five, I promised Admiral Seiberlich I'd write the story of the *Apollo 11* recovery effort from the Navy's perspective so the world would understand more about what happened in the middle of the Pacific in July 1969.

I hope this book meets his expectations.

Bob Fish
Trustee & Apollo Curator
USS Hornet Museum

Hornet *plus* Three

I believe that this nation should commit itself to achieving the goal, before this decade is out, of landing a man on the Moon and returning him safely to the Earth. No single space project in this period will be more impressive to mankind, or more important for the long-range exploration of space; and none will be so difficult or expensive to accomplish.
President John F. Kennedy, in a speech before Congress, May 25, 1961

We choose to go to the Moon. We choose to go to the Moon in this decade and do the other things, not because they are easy, but because they are hard, because that goal will serve to organize and measure the best of our energies and skills, because that challenge is one that we are willing to accept, one we are unwilling to postpone, and one which we intend to win . . . .
President John F. Kennedy's speech at Rice University in Houston, Texas, September 12, 1962

President of the United States John Fitzgerald Kennedy, 1961-1963. Portrait photograph distributed by the White House. Courtesy John F. Kennedy Presidential Library and Museum, Boston.

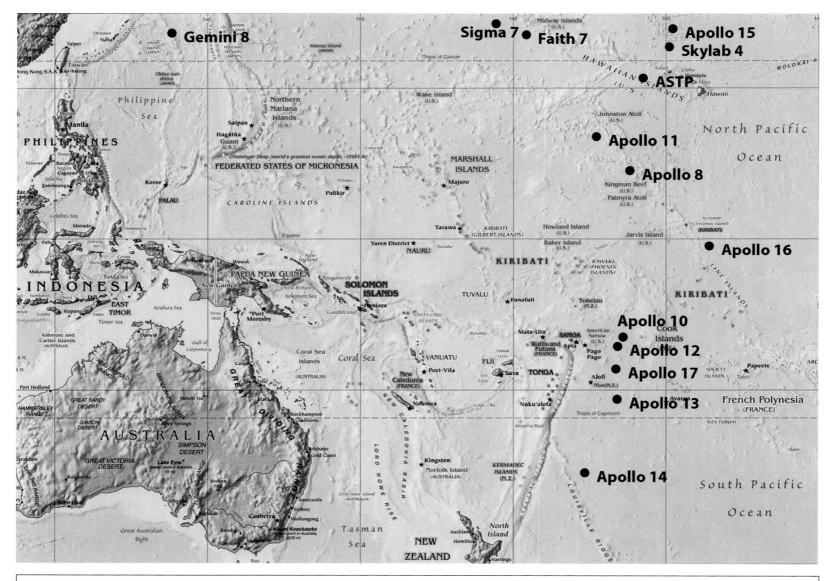

Fig 1A—Map of the spacecraft splashdown and recovery locations in the Pacific Ocean.

Fig 1B—The NASA "meatball" logo, used until 1975

Chapter One
Spacecraft Recovery Program

Many aspects of *Apollo 11*, the first lunar landing mission, have been well documented in news articles, TV documentaries, radio programs and web-based commentaries. With very few exceptions, this information focuses on the space flight itself, covering the period from the Saturn V rocket launch, the three-day flights to and from the Moon, and the few hours spent on the lunar surface by astronauts Neil Armstrong and Buzz Aldrin.

The public record is virtually silent about the splashdown and recovery activities that occurred on July 24, 1969 in the South Pacific. Although over 500 million TV viewers worldwide watched the retrieval process—and many more listened via radio—most only remember the short commentary from President Nixon as he welcomed these space voyagers back to Earth.

Much of this can be traced to disparate public relations efforts between NASA and the Department of Defense (DoD). While humans had sailed ships on the sea for millennia, and had flown aircraft in the sky for a century, they had never walked on another heavenly body before and global interest was intense. Additionally, NASA was a civilian agency, and not in direct line for national defense funding, so they had to keep fueling the public's interest in lunar exploration, which in turn would keep the taxpayer dollars flowing into their coffers. They tasked an elaborate Public Relations machine with keeping their accomplishments in the public eye—and ensuring that most nations were aware of the technological prowess of our democratic system.

The DoD, on the other hand, viewed its support role of the manned space flight program—from Mercury through Skylab—as an important, but secondary, effort to their primary mission of maintaining national security. The Cuban Missile Crisis, war in Vietnam, the *Pueblo* incident, and various hostilities in the Middle East simply took precedence. During the Apollo lunar landing missions, for instance, all of the principal Navy recovery forces in the Pacific had been directly involved with the Vietnam conflict either before or after performing their space flight recovery activities—and sometimes both!

Where NASA celebrated each major achievement very publicly, with ticker-tape parades in New York City and visits to the White House by returning astronauts, the DoD was more interested in maintaining tight security. Giving the media advanced information about ship, air group and personnel deployments was considered tantamount to aiding our enemies. The Navy released a minimum amount of information immediately before, during and after a space mission and then went silent until the next flight. Much DoD information was classified and never released to the public.

With the launch pad disaster of *Apollo 1* in 1967, and the lengthy delay in launch activity that followed, the DoD realized the space mission windows were extremely elastic. Throughout the Moon landing program, launch delays were as commonplace as changes in the weather. President Kennedy's timeframe for placing the first human on the Moon was stated simply as "before this decade is out." For the DoD, the commitment of resources to the combat situation in Vietnam had a very specific timetable and duty cycle. Complex naval resources, such as aircraft carriers with their combat air squadrons, were programmed for deployment to Vietnam, or refit in a shipyard, or decommissioning, more than a year in advance. The massive flow of men, material and machines between the West Coast of the U.S. and the east coast of Vietnam had its own pervasive rhythm. It would have been highly disruptive to interrupt this multi-year cycle, even to support a historic space mission. Once NASA assigned a launch window, the Navy reviewed its available assets and determined which ones could be placed on temporary support duty for the period of the spaceflight.

The relationship between the DoD and the civilian agency NASA was complex, yet symbiotic. In many cases, they had similar goals with regards to access and usage of space, especially Earth orbit activities. In other instances, they were competing for the same funding for research projects. In the end, NASA was highly dependent on DoD support for space flight operations, however, the U.S. government was determined to keep the lunar space program under civilian control.

One of the most fortuitous decisions made by NASA was to utilize passive water-based landings, unlike the Soviets. NASA had several reasons for doing this. The U.S. Air Force missile launch site at Cape Canaveral was originally designed to launch military rockets over the ocean. Thus, although NASA used its own launch complex at the Cape, all missions that experienced an abort-level problem would also have to land in the water, no matter what the "preferred" landing method was.

Another factor was NASA's need to meet a series of incredibly short timetables to achieve a manned lunar landing within the decade. In essence, the agency was barely learning to walk before needing to run at full sprint. Simplicity in engineering provided a higher probability of success. To slow a ground-landing vehicle down adequately required braking rockets, rather than just parachutes, ensuring a more complex spacecraft. For reasons of spacecraft structural design, a water-landing craft weighed much less than one designed to survive a land impact. When all these things were considered, less rocket power was required to launch a water-landing craft into lunar orbit than a land-recovery design. In the end, the Soviet's most powerful rocket, the N1, simply never matured enough to complete a successful launch, while NASA's Saturn V put twelve Americans on the Moon's surface.

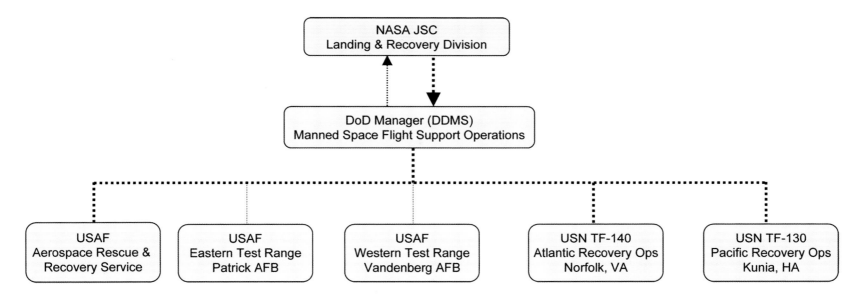

Fig 1C—For spacecraft recovery operations, the Johnson Space Center (JSC) Landing and Recovery Division provided training and mission requirements to the DoD Manager, with the bulk of the assignments going to the Navy task forces.

Department of Defense Support

The Department of Defense played a crucial role in the American space program from the very beginning. In 1958, the office of DoD Manager for Manned Space Flight Support Office (DDMS) was formed with the express purpose of providing effective support to NASA's manned space flight effort. This staff worked closely with various divisions of the NASA Flight Directorate at Johnson Space Center to create and execute optimum support procedures. This support included launch support, worldwide orbital communications, tracking and data relay, a few joint research and development activities, public affairs, and medical support, as well as astronaut and spacecraft recovery operations.

The DoD Manager exercised overall coordination of its worldwide forces based on NASA supplied requirements for that particular mission, whether manned or unmanned. For instance, just prior to an Apollo launch, support forces from five main organizations were assigned to the DoD Manager.

DDMS Recovery Support Overview

During a mission, the deployed Air Force and Navy assets were in direct communication with the DDMS staff located in the NASA Mission Control Center in Houston, Texas. This allowed force commanders in the air or at sea to make timely and effective decisions about recovery efforts in concert with key NASA personnel and based on the latest flight information. DoD land tracking stations, aircraft and ships joined the NASA Manned Space Flight Network stations during Apollo missions to form a global tracking, communication and instrumentation system. The level of responsibilities assumed by each organization varied with the stage of a flight. For instance, the Air Force Eastern Test Range (AFETR) was heavily involved with launch activities, but much less so for the recovery event when Navy units provided the majority of the support.

NASA divided the recovery operations environment into three categories of landing areas—primary, secondary and contingency. The Navy assigned its main recovery assets to the primary landing area, the planned End-Of-Mission (EOM) site that assumed a problem-free flight. When a serious issue arose that prevented a flight from reaching the primary site, one of the secondary landing sites manned by an individual ship would be chosen. The contingency sites would be used only in case of a dire emergency and were covered by USAF aircraft flying from nearby bases.

Five major phases of recovery operations were identified for manned space missions:

- Planning, preparation and training
- Locating the spacecraft
- Providing immediate assistance at the landing site
- Retrieving the flight crew and spacecraft
- Performing post-flight procedures

The combined assets of the DoD and NASA created a vast worldwide logistics capability that ensured everything from life rafts to news media facilities was ready when a space flight occurred. Flexibility was a key ingredient to success—weather influences, minor launch delays, or unforeseen emergencies had to be accommodated. As units were designated for assignment to a space mission, they underwent a programmed training cycle. NASA created a variety of special equipment to ensure effective recovery operations and these items were deployed to all potential recovery forces well before the launch.

USAF Major General David M. Jones was the DoD Manager for DDMS during the *Apollo 11* mission, in addition to commanding AFETR.

U.S. Air Force Overview

Due to their early rocket development and flight activities, AFETR had an extensive monitoring and tracking network as well as a missile launch complex and support capability at Cape Canaveral. Their sea and land-based range monitoring capabilities extended from Florida through the south Atlantic and into the Indian Ocean. The Air Force Western Test Range (AFWTR) handled similar functions for the missile test ranges in the Pacific, extending from California to Eniwetok in the Marshall Islands.

The USAF also had the Aerospace Rescue and Recovery Service (ARRS), which provided emergency rescue operations capability around the world during Earth orbit missions. Up to thirty aircraft were deployed for a space mission, covering the expansive ocean areas between Navy ships. A spacecraft contingency landing could become necessary at any moment, from launch to mid-flight to planned landing. Therefore, just prior to a mission launch, ARRS aircraft would be dispatched to pre-determined contingency landing sites prepared to drop pararescue jumpers and remain in orbit over the spacecraft as long as necessary to stabilize the spacecraft and crew.

U.S. Navy Overview

The Navy augmented the USAF range monitoring functions with specially instrumented ships. However, its primary role was to ensure the availability of well-prepared and pre-positioned spacecraft recovery forces. Rigorous training was provided for recovery teams, modifications made to potential recovery vessels, and multiple ships were dedicated to each space mission. Recovery crews were trained for various contingencies, ships properly equipped, medical personnel embarked, and all other on-site logistical services were provided to ensure a successful recovery. The Navy always assigned an aircraft carrier or amphibious (helicopter) assault ship as Primary Recovery Ship (PRS). Essentially floating cities, they could remain at sea for long periods of time if issues arose that required extending a mission. They had significant medical capability onboard along with the ability to evacuate injured astronauts by plane or helicopter. Finally, there was plenty of room for all the press and dignitaries who accompany a huge media event like a Moon landing. Destroyers were usually selected as the secondary recovery ships.

Task Force 140

In 1961, the Navy's Atlantic Command created TF-140 to support Project Mercury, coordinating Atlantic spacecraft recovery operations from Norfolk, Virginia. TF-140 was primarily a staff organization, reporting to the DDMS office. There was no standing force, but rather a contractual arrangement to temporarily command ships, aircraft squadrons, and rescue swimmers who were trained and equipped for recovery missions on an as-needed basis. As new ships were assigned, spacecraft recovery training exercises were conducted, often in the Virginia Capes area. During the *Apollo 11* mission, Rear Admiral Philip S. McManus was its commander.

Task Force 130

In 1962, when the NASA space program entered its Earth-orbiting stage, the Navy created TF-130 to manage the Pacific theater of recovery operations. The commander and his staff of ten officers and five enlisted personnel coordinated these efforts from their center at Kunia, Hawaii on the island of Oahu. During the *Apollo 11* mission, Rear Admiral Donald C. Davis was the commander of TF-130 and Captain Robert T. Tolleson was the Recovery Officer on his staff.

Summary

Although the manned missions were better known, Air Force and Navy forces were also involved in the recovery of numerous unmanned flights that gathered valuable scientific information as well as testing spacecraft components in a "production" environment. For instance, USS *Hornet* recovered the unmanned AS-202 flight in August 1966 near Wake Island, a mission that tested the ability of the spacecraft's ablative heat shield to withstand the high temperatures expected during a lunar mission re-entry into the Earth's atmosphere.

Overall, the DoD provided launch, flight and recovery support for thirty-one manned missions between 1961 and 1975, employing well over 200 different units. Table 1 provides a complete list of the manned spacecraft recoveries performed by the Navy. All but two Mercury and one Gemini missions splashed down in the Atlantic as well as both Earth-orbit Apollo flights. All the lunar orbit and lunar landing fights landed in the Pacific.

The Moon race was a key component of the 1960's struggle for world supremacy between Democracy and Communism. Given the national strategic imperative behind the lunar program, the DoD did an outstanding job of supporting the American race to the Moon.

Fig 1D—The logo of Navy TF-140 responsible for coordinating spacecraft recovery operations in the Atlantic Ocean.

Fig 1E—The logo of Navy TF-130 responsible for coordinating spacecraft recovery operations in the Pacific Ocean.

Fig 1F—During its summer NROTC cruise, *Hornet* recovered the Block 1 Apollo Command Module from NASA's unmanned AS-202 mission on August 25, 1966. The spacecraft was recovered about 400 miles east of Wake Island.

Table 1 – List of Manned Spacecraft Recoveries

NASA MISSION	FLIGHT CREW	NAVY VESSEL	OCEAN	DATE
Mercury - *Freedom 7*	Shepard	USS *Lake Champlain* (CVS-39)	Atlantic	05/05/61
Mercury – *Liberty Bell 7*	Grissom	USS *Randolph* (CVS-15)	Atlantic	07/21/61
Mercury – *Friendship 7*	Glenn	USS *Noa* (DD-841)	Atlantic	02/20/62
Mercury – *Aurora 7*	Carpenter	USS *John R. Pierce* (DD-753)	Atlantic	05/24/62
Mercury – *Sigma 7*	Schirra	USS *Kearsarge* (CVS-33)	Pacific	10/03/62
Mercury – *Faith 7*	Cooper	USS *Kearsarge* (CVS-33)	Pacific	05/15/63
Gemini 3	Grissom, Young	USS *Intrepid* (CVS-11)	Atlantic	03/23/65
Gemini 4	McDivitt, White	USS *Wasp* (CVS-18)	Atlantic	06/07/65
Gemini 5	Cooper, Conrad	USS *Lake Champlain* (CVS-39)	Atlantic	08/29/65
Gemini 6A	Schirra, Stafford	USS *Wasp* (CVS-18)	Atlantic	12/16/65
Gemini 7	Borman, Lovell	USS *Wasp* (CVS-18)	Atlantic	12/18/65
Gemini 8	Armstrong, Scott	USS *Leonard F. Mason* (DD-852)	Pacific	03/17/66
Gemini 9	Stafford, Cernan	USS *Wasp* (CVS-18)	Atlantic	06/06/66
Gemini 10	Young, Collins	USS *Guadalcanal* (LPH-7)	Atlantic	07/21/66
Gemini 11	Conrad, Gordon	USS *Guam* (LPH-9)	Atlantic	09/15/66
Gemini 12	Lovell, Aldrin	USS *Wasp* (CVS-18)	Atlantic	11/15/66
Apollo 7	Schirra, Eisele, Cunningham	USS *Essex* (CVS-11)	Atlantic	10/22/68
Apollo 8	Borman, Lovell, Anders	USS *Yorktown* (CVS-10)	Pacific	12/27/68
Apollo 9	McDivitt, Scott, Schweickart	USS *Guadalcanal* (LPH-7)	Atlantic	03/13/69
Apollo 10	Stafford, Young, Cernan	USS *Princeton* (LPH-5)	Pacific	05/26/69
Apollo 11	Armstrong, Collins, Aldrin	USS *Hornet* (CVS-12)	Pacific	07/24/69
Apollo 12	Conrad, Gordon, Bean	USS *Hornet* CVS-12)	Pacific	11/24/69
Apollo 13	Lovell, Swigert, Haise	USS *Iwo Jima* (LPH-2)	Pacific	04/17/70
Apollo 14	Shepard, Roosa, Mitchell	USS *New Orleans* (LPH-11)	Pacific	02/09/71
Apollo 15	Scott, Worden, Irwin	USS *Okinawa* (LPH-3)	Pacific	08/07/71
Apollo 16	Young, Mattingly, Duke	USS *Ticonderoga* (CVS-14)	Pacific	04/27/72
Apollo 17	Cernan, Evans, Schmitt	USS *Ticonderoga* (CVS-14)	Pacific	12/19/72
Skylab 2	Conrad, Weitz, Kerwin	USS *Ticonderoga* (CVS-14)	Pacific	06/22/73
Skylab 3	Bean, Lousma, Garriott	USS *New Orleans* (LPH-11)	Pacific	09/25/73
Skylab 4	Carr, Pogue, Gibson	USS *New Orleans* (LPH-11)	Pacific	020/8/74
Apollo – Soyuz	Stafford, Slayton, Brand	USS *New Orleans* (LPH-11)	Pacific	07/24/75

Table 1 —Of the 31 manned missions, all of the astronauts, and all but one spacecraft (*Liberty Bell 7*), were retrieved by Navy ships. USAF Pararescuemen participated in the recovery of *Aurora 7* and *Gemini 8*. Helicopters from USS *Intrepid* (CVS-11) recovered astronaut Carpenter from *Aurora 7*.

Fig 2A—Astronaut Alan Shepard nears the helicopter hatch after his successful retrieval from the *Freedom 7* capsule.

Chapter Two
The Learning Curve

Initial Preparations

The DoD's support forces and recovery techniques evolved with the constantly upgrading NASA capabilities. The first two manned Mercury flights in 1961 were sub-orbital in nature, requiring coverage only in the Caribbean area, and were easily handled by a few ships and two helicopter groups. During the final orbital Mercury flight (MA-9) in 1963, NASA designated over thirty-two possible landing sites around the world (plus launch abort sites), which the USAF and the Navy covered with 171 aircraft and twenty-eight surface ships!

The headquarters for NASA's Mercury program was at the Langley Research Center in Hampton, Virginia. As a result, Marine Corps helicopter squadrons and Navy Underwater Demolition Teams stationed nearby were called upon to help develop astronaut survival and recovery procedures.

Because the medium cargo-transport helicopter units at Marine Corps Air Station (MCAS) New River, North Carolina worked through these initial exercises with NASA, they were assigned to perform the recoveries of early Mercury space flights. They flew from the Navy's Primary Recovery Ship (PRS) and were backed up by the ship's helicopters.

The single-person Mercury spacecraft was relatively light, so the initial recovery process called for using just a single recovery helicopter. The helicopter hovered about thirty feet above the capsule as it bobbed in the water. A cable was attached to a strong-point on the top of the capsule and winched tight to stabilize the spacecraft, and by raising it slightly, allowed water in the landing bag to drain. Then a horse collar sling was lowered to the astronaut who was winched up next to the cargo hatch and assisted into the helicopter. Finally, the helicopter lifted the capsule free of the water and everything was flown back to the PRS for post-flight examination.

For Marine units, the medium lift helicopter of choice was the Sikorsky H-34 Seahorse (also known as a HUS-1). As the role of the Marines in the Vietnam conflict expanded, and updated NASA spacecraft exceeded the limited lifting capability of the Seahorse helicopter, Marine support of the NASA space program was phased out when Project Mercury ended.

In the early 1960's, the Navy's Underwater Demolition Teams (UDT) were headquartered at two Naval Amphibious Bases (NAB) in the U.S. Atlantic Ocean operations were handled by NAB Little Creek near Norfolk, Virginia while Pacific Ocean responsibilities were carried out by NAB Coronado near San Diego, California. Each UDT team had a two-digit identifier, such as UDT-11. The designation for teams stationed at Little Creek began with a "2" while those based in Coronado began with a "1."

The UDT team at Little Creek was involved from the very beginning of the manned space program, providing water egress training for the original Mercury 7 astronaut corps. They helped NASA's Landing and Recovery Division design special equipment, document survival techniques, and create training programs to ensure the safety of its astronaut corps once a landing had occurred. UDT personnel remained closely involved with the NASA recovery efforts until the Space Shuttle program. UDT-21 handled the Mercury and Gemini mission recoveries in the Atlantic Ocean while UDT-22 did the Earth-orbit *Apollo 9* mission. The West Coast teams based at Coronado rotated the recovery assignments among three teams, UDT-11, -12 and -13, depending on the combat situation in Vietnam. Two Mercury flights, *Sigma 7* and *Faith 7*, landed in the Pa-

Fig 2B—The picture-perfect recovery of *Freedom 7* (MR-3). Astronaut Alan Shepard is hoisted up to the USMC Seahorse helicopter. *Freedom 7* is attached by another cable, seen here stretched taut as the landing bag is allowed to drain its sea water.

cific and were recovered with assistance from UDT-11 and -12. All Apollo lunar orbit and landing missions splashed down in the Pacific.

Project Mercury

On May 5, 1961, the recovery of the first U.S. manned space flight, *Freedom 7* (MR-3), a sub-orbital mission flown by astronaut Alan Shepard, went according to plan. The primary recovery ship USS *Lake Champlain* (CVS-39) was perfectly positioned 300 miles downrange (southeast) from Cape Canaveral before the launch. Three USMC helicopters from Marine Air Group 26 (MAG-26) and two Navy helicopters from Helicopter Squadron 5 (HS-5) visually followed the descent of the spacecraft under its single parachute. They were in the immediate area when *Freedom 7* splashed down only six miles from the ship.

Lieutenant Wayne Koons, pilot of Helicopter #44, hovered low over *Freedom 7*, as copilot Lieutenant George Cox latched the special NASA shepherd's hook onto the recovery loop on top of the capsule. That cable was pulled taught to stabilize the capsule.

Shepard exited through the top hatch and placed the "horse collar" harness around himself. He was winched up to the open cargo door and helped into the helicopter by the copilot. Once the water had drained from the spacecraft's landing bag, both astronaut and capsule were flown back to the ship. As the helicopter approached *Lake Champlain*, Shepard told Cox "This is one of the best carrier landings I've ever made." After the capsule was lowered onto a special pad, the H-34 landed and Shepard jumped onto the flight deck. The astronaut soon received a congratulatory phone call from President Kennedy.

However, only two months later on July 16, the recovery operation on the second manned flight, *Liberty Bell 7* (MR-4), barely averted total disaster. The flight was fine and USS *Randolph* (CVS-15) was on station with its helicopters circling the descending spacecraft until it splashed down. After the capsule landed safely in the water, Marine pilot Lieutenant Jim Lewis flew his Seahorse (#32) toward the bobbing capsule and ra-

Fig 2C—USMC pilot Jim Lewis dragged *Liberty Bell 7* away from drowning astronaut Gus Grissom whose head can be seen in the water beneath the rear wheel. Lewis was unable to drain enough water from the capsule to lift it completely and it sank moments later, nearly taking the helicopter with it.

dioed astronaut Gus Grissom to signal when he was ready. A few minutes later, the helicopter approached *Liberty Bell 7* to allow copilot Lieutenant John Reinhard to cut the radio whip antenna and lower the hook to latch onto the recovery loop. Just before the helicopter arrived overhead, explosive bolts on the side hatch of *Liberty Bell 7* detonated prematurely, blowing off the hatch cover. As water rushed in, Grissom hurriedly exited the spacecraft and swam a short distance away, but he was unable to shut a hose inlet valve in his spacesuit nor grab a life raft. Within a minute, both the capsule and astronaut's spacesuit were filling with water. The Mercury capsule started to sink beneath the waves, so the Seahorse latched on to it even as the helicopter's wheels dipped into the sea. Although Lewis was able to get the capsule back to the surface, the amount of water inside made it too heavy for the Seahorse to lift completely free of the ocean. It would have been dangerous to attempt to retrieve Grissom as well.

Grasping the gravity of the personnel situation, Lewis recalled, "I dragged the spacecraft away from the struggling astronaut. Whether I could save the capsule or not, we certainly wanted to get Grissom out of the water." Because Lewis had moved the capsule away from the astronaut, a backup helicopter (#30) was able to swoop in. Its copilot, Lieutenant George Cox (who had recovered Alan Shepard), lowered the horse collar sling to Grissom. Barely able to keep his head above water, he struggled to get the rescue harness around himself. He finally got it on, although backwards, and was lifted into the second Seahorse for the short flight to *Randolph*.

Meanwhile, Lewis battled on, trying to drag the capsule back to the recovery ship so its boat and aircraft (B&A) crane could finally make the retrieval. The water-logged capsule weighed 1,000 pounds more than the Seahorse's rated lifting capacity. The H-34's engine was over-taxed and a red light soon appeared on the pilot's panel indicating a possible engine failure. Lewis knew *Liberty Bell 7* had to be released, and did so. Very quickly, it sank to the ocean floor 15,000 feet below, where it remained for thirty-eight years.

NASA engineers designed the Mercury spacecraft primarily for space flight efficiency, not for floating on water, as highlighted by this incident. Both NASA and the Navy realized the margin for error in the original procedures was far too slim and started working on a more effective spacecraft recovery procedure.

Under the guidance of Don Stullken, NASA's Landing and Recovery Division quickly completed the development of an auxiliary flotation collar that could be attached to the spacecraft once it attained a right-side up (Stable 1) position. In addition to providing buoyancy, the collar would help stabilize the spacecraft's bouncing motion, caused by ocean swells, and act as a work platform for personnel at the scene, whether medical, rescue or astronaut. For all future missions, this device was applied by military rescue divers immediately after splashdown, rather than leaving the astronaut(s) bobbing in a cramped spacecraft with dubious floating properties.

Before the Mercury program ended, USAF and Navy special operations personnel were routinely practicing their jumps from aircraft of various types into the sea near wallowing space capsules, attaching a flotation collar and a life raft.

In May 1962, the expanding capabilities of the DoD support effort proved their value. Due to a late firing of the retro-fire control system, Scott Carpenter and *Aurora 7* [MA-7] landed 250 miles downrange from the primary recovery ship USS *Intrepid* (CVS-11). Carpenter, knowing there would be a wait for the recovery forces to arrive, exited the capsule and floated in a one-man life raft. Using radio direction-finding equipment, a USAF ARRS SC-54 aircraft (military version of the DC-4) sped to the scene. Two pararescue jumpers from the 41st Air Rescue Squadron parachuted into the water from 1,000 feet overhead. They ensured the astronaut was safe and then secured the capsule with a flotation collar.

John Stonesifer, the NASA Recovery Team Leader recalls, "Due to the distance involved, *Intrepid* elected to use the brand new SH-3A Sea King helicopters to airlift the recovery team to the site and retrieve the astronaut." It was the first use of a Navy Sea King, which soon became the rotary-wing aircraft of choice for both Gemini and Apollo.

About three hours after landing, Carpenter returned to the carrier while the spacecraft was winched aboard the destroyer USS *John R Pierce* (DD-753) and delivered to Puerto Rico.

The final two Mercury missions, MA-8 and MA-9, landed in the Pacific Ocean. After splashdown, the astronauts elected to remain inside the capsule. For safety reasons, the ships did not directly approach the capsule for pickup, but rather sent their motor launch out with a towing line or fired a shot line to the UDT swimmers. Once the UDT swimmers connected the line to the spacecraft, it was then slowly "reeled in" until under the B&A crane. It was hoisted aboard and the astronaut assisted out of the spacecraft.

This solution of astronauts remaining in the spacecraft until being hoisted aboard the PRS would not be practical for most of the missions that followed. There was intense debate in NASA about whether to leave the astronauts inside, or retrieve them via helicopter, before hoisting a spacecraft onto the ship. Considerations included:

- The unpredictability of the sea state at the splashdown point (wind, waves, drift rates, etc)
- The differences in seagoing dynamics between a giant aircraft carrier and a tiny spacecraft
- The need to conduct speedy physical exams on the astronauts following their lengthy stay in a zero gravity environment.

Fig 2D—UDT swimmers have attached a flotation collar to the spacecraft *Faith 7* (MA-9). A lifeboat from USS *Kearsarge* (CVS-33) rowed a towing line out to the swimmers. Astronaut Gordon Cooper remained inside while the capsule was pulled to the side of the recovery ship.

Fig 2E—The USS *Kearsarge* (CVS-33) B&A crane has lifted the MA-9 spacecraft, with astronaut Cooper inside, onto elevator #3. Navy personnel hold the capsule steady while the astronaut (in white flight suit) is assisted out through the side hatch.

In the end, NASA opted to use helicopter retrieval of astronauts immediately following splashdown. They felt the risks were greatly lessened by using Navy flight crews who were highly trained to make pickups of personnel from the water and could reach the landing area more quickly than a ship. Unfortunately, this decision would add major complexity to the recoveries of the initial Apollo lunar landing flights because of the "Moon germ" issue.

Project Gemini

When NASA moved on to the two-man Gemini spacecraft, the Navy significantly upgraded its recovery capabilities and technology as well. The Gemini capsule was designed to float on its side. Its increased weight over the Mercury spacecraft made it too heavy to be safely lifted by any helicopter in the Navy's inventory. For most of the remaining Gemini and Apollo space programs, astronauts were hoisted into the recovery helicopter while the spacecraft awaited the arrival of the ship and its heavy lifting B&A crane. As a result, UDT divers now also attached a sea anchor and one or more life rafts to the bobbing spacecraft in addition to the flotation collar. This allowed the astronauts to egress the capsule safely without entering the water, and also provided a platform from which the helicopters would lift the astronauts. The Navy decided to use the more powerful, and all-weather capable, Sikorsky SH-3 Sea King helicopter configured in its search and rescue (SAR) mode. While the helicopter hovered just to the side and forty feet above the spacecraft, the astronauts were raised one at a time into the cargo bay and then flown back to the PRS.

The *Gemini 4* mission provided a good demonstration of the new recovery procedures. After splashing down under its two parachutes, the spacecraft floated on its side. Green dye was automatically released into the water to provide a highly visible marker of the capsule location for helicopter pilots. UDT divers jumped out of their Sea King and installed a flotation collar around the spacecraft, tethered a raft to it and assisted the astronauts into the raft. The helicopter moved into position, hovering overhead so the aircrewmen could winch the astronauts out of their raft, one at a time. Two UDT divers were positioned in the water near the "hoist raft" to rescue either astronaut should an unforeseen

Fig 2F—Overhead view of the *Gemini 4* spacecraft showing the yellow flotation collar used to stabilize the spacecraft. The green marker dye is highly visible from the air and is used as a locating aid. A crewmember is being hoisted up to a Sea King helicopter following the successful four-day mission.

event occur during the lift operation. Another diver closed the hatches on the Gemini capsule, preparing it for the B&A crane lift onto the carrier's elevator. Once astronauts Edward White and James McDivitt were safely in the helicopter, they were flown back to USS *Wasp* (CVS-18).

This August 1965 photograph shows the recovery of *Gemini 5* once the astronauts, Gordon Cooper and Pete Conrad, were retrieved by a helicopter. The USS *Lake Champlain* (CVS-39) very slowly crept toward the capsule until it was abeam of the deck-edge elevator and about fifty yards out. An inhaul line was provided to the UDT team on top of the spacecraft and it was pulled over to the side of the ship. A special NASA "hook assembly" attached to the B&A crane was connected to the capsule lifting device. The crane operator skillfully lifted the capsule free of the water. He swung it over to the aircraft elevator and lowered it to within a few feet of the deck so the ship's crew could remove the flotation collar. It was then raised back up a little and placed onto a special dolly with four wheels. The capsule was towed into the hangar bay for initial post-retrieval procedure. These included the removal of onboard data and material, recording of instrument readings and switch positions and a preliminary examination of its physical condition.

Over the course of the Gemini program, the most unusual recovery was of *Gemini 8*. A stuck thruster on the capsule sent it into a spin while in orbit, forcing its astronauts, Neil Armstrong and David Scott, to abort their mission early. Showing great skill, they flew the spacecraft manually to a designated secondary landing area in the East China Sea. The secondary landing area ship, the destroyer USS *Leonard F. Mason* (DD-852), was 180 miles from the probable *Gemini 8* landing site when the decision was made to terminate the flight. Steaming at flank speed, *Mason* was still three hours away from the spacecraft when it splashed down 400 miles southeast of Okinawa, the other side of the world from its planned landing site.

An ARRS HC-54 Rescuemaster on standby in Okinawa, radio call-sign "*NAHA Rescue One*," was alerted to the emergency situation as it was developing. Vectored to the general splashdown location by NASA before the retro-rockets were fired, the pilot spotted the capsule just before it hit the water. Three pararescue jumpers (PJs) bailed out from 1,000 feet overhead. Within minutes of the spacecraft's landing, A/2c Glenn Moore, A/1C Eldridge Neal, and S/SGT Larry Huyett secured a flotation collar to it and provided as-

Fig 2G—The *Gemini 5* spacecraft is hoisted aboard USS *Lake Champlain* by the ship's B&A crane. Navy personnel have just removed the flotation collar so the capsule can be eased down onto a mobile dolly (painted white) for transport. Note the special NASA hook assembly that connects the capsule to the crane.

sistance needed to ensure the survival of the crew until the *Mason* arrived. When the ship reached the scene, the flight crew and PJs were immediately assisted up a Jacob's ladder onto the destroyer's deck and taken to the medical department. Once the capsule was winched aboard using a special NASA crane assembly issued to all secondary recovery ships, everyone headed to Naha, Okinawa.

This unplanned recovery operation clearly demonstrated how thoroughly the DoD and NASA had foreseen various contingencies and how well the appropriate DoD contingency forces were trained and positioned to respond quickly to emergencies.

Failure was simply not an option.

Fig 2H—After completion of their problem-shortened *Gemini 8* mission, astronauts Armstrong and Scott sit with their spacecraft hatches open as the recovery ship USS *Leonard F. Mason* pulls alongside. Moments after this photo was taken from the *Mason*, two seasick astronauts and three queasy PJs were assisted up a "Jacob's rope ladder" onto the deck of the destroyer.

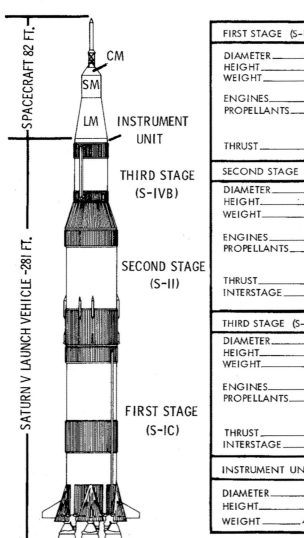

FIRST STAGE (S-IC)	
DIAMETER	33 FEET
HEIGHT	138 FEET
WEIGHT	5,022,674 LBS. FUELED
	288,750 LBS. DRY
ENGINES	FIVE F-1
PROPELLANTS	LIQUID OXYGEN (3,307,855 LBS., 346,372 GALS.) RP-1 (KEROSENE) - (1,426,069 LBS., 212,846 GALS.)
THRUST	7,653,854 LBS. AT LIFTOFF

SECOND STAGE (S-II)	
DIAMETER	33 FEET
HEIGHT	81.5 FEET
WEIGHT	1,059,171 LBS. FUELED
	79,918 LBS. DRY
ENGINES	FIVE J-2
PROPELLANTS	LIQUID OXYGEN (821,022 LBS., 85,973 GALS.) LIQUID HYDROGEN (158,221 LBS., 282,555 GALS.)
THRUST	1,120,216 TO 1,157,707 LBS.
INTERSTAGE	1,353 (SMALL)
	8,750 (LARGE)

THIRD STAGE (S-IVB)	
DIAMETER	21.7 FEET
HEIGHT	58.3 FEET
WEIGHT	260,523 LBS. FUELED
	25,000 LBS. DRY
ENGINES	ONE J-2
PROPELLANTS	LIQUID OXYGEN (192,023 LBS., 20,107 GALS.) LIQUID HYDROGEN (43,500 LBS., 77,680 GALS.)
THRUST	178,161 TO 203,779 LBS.
INTERSTAGE	8,081 LBS.

INSTRUMENT UNIT	
DIAMETER	21.7 FEET
HEIGHT	3 FEET
WEIGHT	4,306 LBS.

Fig 3A—This NASA diagram shows the 363-foot high Apollo "stack" as it would look on the launch pad. The S-IC (first stage) rocket accelerated the spacecraft to 6000 miles per hour. The S-II second stage increased that to 15,300 miles per hour. Finally, the single engine S-IVB third stage took over, putting the spacecraft into orbit. Its engine was designed to be restarted in order to make the Trans-Lunar Injection burn.

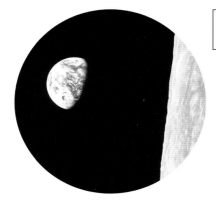

Fig 3B—Dramatic photo of Earthrise taken from *Apollo 8* after lunar orbit insertion burn.

Chapter Three
DoD Apollo Mission Support

Apollo Command Module Recovery Concept

During the launch, an Apollo spacecraft consisted of two major components, the Command and Service Module (CSM) and the Lunar Module (LM). The CSM had two main components: the Service Module (SM) and the Command Module (CM). The Service Module provided the primary propulsion and maneuvering capability of the spacecraft. Most of the consumables (oxygen, hydrogen and propellant) were stored in this module, which was jettisoned just before re-entry into the Earth's atmosphere. The most publicly recognizable part of the CSM was the Command Module, which was shaped like a blunt cone. The Command Module was roughly ten feet in height, twelve feet in diameter and weighed approximately 10,000 pounds upon landing. With three adjustable couches and five windows, it served as the crew compartment and control center. It was equipped with a heat shield at the bottom of the cone and an Earth Landing System (ELS) located in the top (i.e., apex).

The ELS primarily consisted of three large main parachutes, three pilot parachutes, two drogue parachutes, and a recovery aid subsystem. The drogue and pilot parachutes were mortar-deployed to ensure they were ejected beyond the turbulent air following the Command Module as it fell through the sky. Unlike the Mercury and Gemini spacecraft, the five-ton Apollo CM landed with three orange and white ring-sail parachutes, each having a diameter of eighty-three feet and suspension lines 160 feet long. They were designed to reduce the descent rate of the CM to just twenty-two miles per hour and position the spacecraft at a twenty-seven degree pitch angle to hit the water on its "toe" for the least shock to the astronauts.

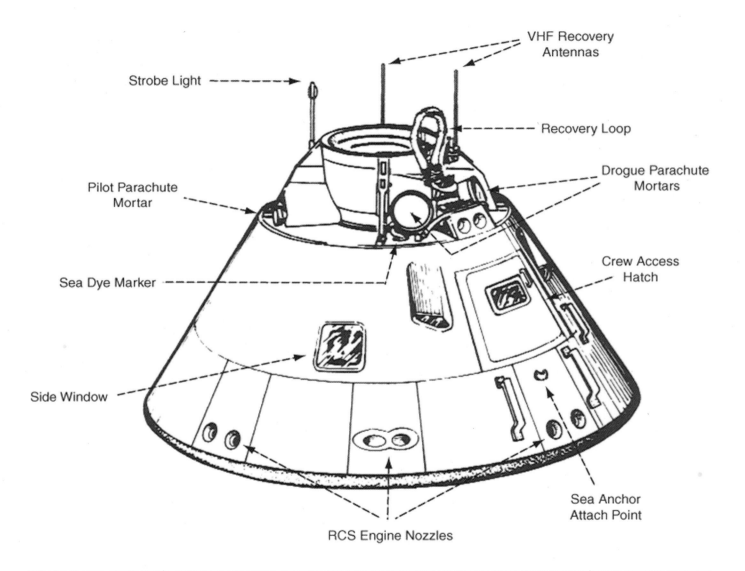

Fig 3C—This Apollo Command Module graphic shows many of the aids used for locating and recovering the spacecraft after splashdown.

After splashdown, the main parachutes were released and the recovery aid subsystem was activated by the crew. The subsystem consisted of three uprighting bags, a swimmer's umbilical cable, a dye marker canister, a flashing beacon and a VHF beacon transmitter. A recovery sling made of steel cable also was provided to lift the CM aboard ship. The twelve-foot long swimmer's umbilical allowed the first frogman on the scene to connect his communications equipment so he could talk to the astronauts without opening the hatch.

The Command Module was the only part of the spacecraft that returned to Earth and was the focus of the DoD recovery effort.

NASA created pre-production spacecraft—called boilerplates—that were similar to their production counterparts in size, shape, mass and center of gravity. Boilerplates, and occasionally an actual spacecraft, were used to conduct parachute research and development, water drop tests, studies of stability characteristics, vibration tests, flight tests, and for other purposes helpful for the proper design and development of the actual spacecraft and its systems. Some of these boilerplates were equipped with instrumentation to permit recording of data for engineering study, while others were just made with a metal skin (i.e., made out of thick steel normally used to make boilers). NASA provided a number of metal boilerplates to the Navy so their recovery units could practice techniques without endangering a multi-million dollar spacecraft in the process.

Final DoD Preparations

By the time NASA was ready to send a manned mission to the Moon, the framework for recovering Earth-orbit missions had been thoroughly worked out and rehearsed. But now they had to prepare for any possible contingencies that might occur during a deep space flight. Due to the natural dynamics of flight geometry for achieving lunar orbit, there were two launch windows every day. Also, to ensure a full systems checkout, Apollo lunar flights had to orbit the Earth at least once before beginning their three-day journey to the Moon. For a daytime launch in July 1969, any space flight destined to orbit the Moon would enter its trans-lunar flight path over the middle of the Pacific Ocean. The event that began this course-change is the Trans-Lunar Injection (TLI) burn of the Saturn S-IVB third-stage engine. The function of the TLI burn was to place the spacecraft in a huge, elongated, elliptical orbit that intercepted the Moon (an orbit with an apogee of about 310,000 miles).

Because a malfunction might occur at any time during the Earth orbits, NASA and the DoD created several "mission abort" as well as "normal recovery" lines around the globe. Ships and/or aircraft were assigned to cover the Atlantic Ocean Line, Indian Ocean Line, Western Pacific Line, Mid-Pacific Line and East Pacific Line. Once the spacecraft completed its TLI engine burn and achieved a lunar orbit trajectory, there were only two possible recovery lines: the Mid-Pacific Line (MPL) and Atlantic Ocean Line (AOL). The preferred End-Of-Mission (EOM) landing site was the Mid-Pacific Line, running north-south a few hundred miles west of Hawaii. Regardless of which recovery line was used, the latitude of the landing remained approximately the same as the latitude at which the TLI event occurred.

It became evident to NASA and the DoD during the Gemini program, that having Navy ships cover every possible contingency landing site was neither reasonable nor cost-effective. Following WWII, the USAF had established bases around the world to handle emergency recovery of crews from aircraft that crashed or were forced to land in remote locations. The unit tasked with performing these rescues, whether on land or sea, was called the Aerospace Rescue and Recovery Service (ARRS). Providing support for the space program was a natural extension to their military mission.

For Apollo contingency recovery operations, ARRS units flew the HC-130H aircraft, which carried special radio direction finding and surface-to-air-recovery equipment. The HC-130H carried an aircrew of eleven: pilot, copilot, navigator, radio operator, two flight mechanics, two loadmasters and three pararescue jumpers. If called into action following an emergency situation, the flight crew had to first locate the spacecraft, then contact the appropriate Navy recovery command center to relay coordinates and finally deploy a pararescue team with a special Apollo support package.

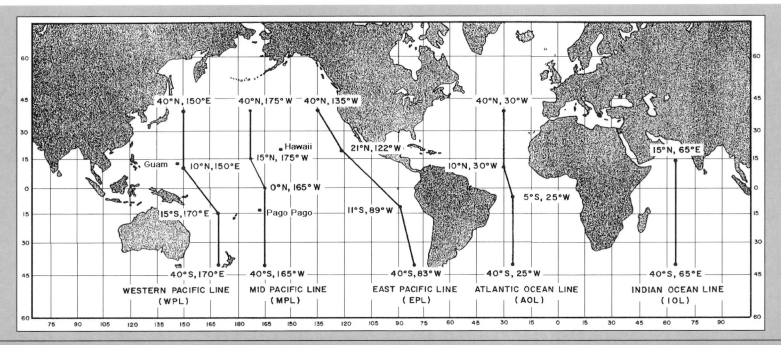

Fig 3D—This mimeographed chart is from the *Apollo 11* Press Kit handed out by TF-140. It shows the five worldwide recovery lines where an Apollo mission could end, either from its Earth orbit or lunar return flight. Once inserted into a lunar orbit trajectory, the MPL line was the preferred target.

Apollo missions required a worldwide network to perform tracking, telemetry and communication functions during the Earth orbit phase. Several land-based stations, previously used for the Mercury and Gemini programs, were scattered around the world with many in remote locations. They performed the requisite monitoring and relay functions for their period of communications visibility with the spacecraft.

Unfortunately, two critical Apollo mission-related events took place in the southwestern Pacific Ocean where there were no land stations. The first one was the Trans-Lunar Injection burn that propelled the spacecraft out of Earth orbit and on its way to the Moon. A week or so later, the second one occurred when the spacecraft re-entered Earth's atmosphere and began its descent for landing in the mid-Pacific. During the Mercury and Gemini programs, these gaps in the communication systems were tolerated. For these key Apollo events, personnel in NASA's Mission Control Center wanted assured voice and telemetry links with the astronauts and spacecraft.

For part of this solution, NASA decided to expand the range instrumentation ship fleet currently serving with the USAF eastern and western missile test ranges. Three World War II era oil tankers were enlarged and outfitted with sophisticated electronic equipment. Known as Apollo Instrumentation Ships (AIS) they became "mobile spacecraft monitoring bases." For the launch, Earth orbit insertion, lunar orbit insertion and End-Of-Mission phases of an Apollo flight, AIS units were stationed in areas where the land-based coverage was inadequate. Whenever a spacecraft was outside the Earth's atmosphere, NASA's Deep Space Network took over all communications. In many cases, however, the AIS fleet would loiter near their launch stations to await the re-entry event.

Before a launch, it was not known on which Earth orbit the TLI burn would take place; the successive orbital paths could be several hundred miles apart. The instrumentation ships were assigned to the most likely (i.e., nominal) position days before a launch, but could not move quickly enough to cover other geographical positions should a variance in the flight plan occur. Since they floated on the sea's surface, they were also unable to photograph events happening in the upper atmosphere and were effectively blind during the re-entry phase when there was a communications blackout with the spacecraft. This meant gaps in the Apollo telemetry coverage would still occur.

Fig 3E—ARIA EC-135 number 330 is shown flying over the Apollo Instrumentation Ship USNS *Vanguard*. The range instrumentation and tracking ships would often be on station for days before a mission while the ARIA aircraft were dispatched only a few hours before the launch.

NASA decided to develop an even more mobile capability to augment the ships. NASA and the USAF jointly developed an airborne solution called the Apollo Range Instrumentation Aircraft (ARIA). Eight militarized Boeing-707 jet cargo aircraft were modified to handle telemetry and communications similar to the AIS units. They flew to their designated position just hours before the mission event, handled their tracking and relay functions and then returned to the nearest air base capable of logistically supporting them. If an Apollo event was delayed by one orbit, the aircraft had ninety minutes to fly to a location closer to its orbital path. For more information about the AIS and ARIA solutions, please refer to the Technical Note at the end of this chapter.

Nominal DoD Mission Deployment

Just prior to the launch of an Apollo flight, the DoD Manager had numerous ships, dozens of aircraft (ARIA, ALOTS, ARRS, etc), Navy UDT and Air Force pararescue teams on station around the world. The TF-130 and TF-140 staffs maintained a twenty-four-hour surveillance of the position and readiness of all recovery forces, and weather conditions in the primary and contingency splashdown areas. The primary and secondary landing site teams underwent extensive training using NASA supplied equipment such as boilerplate command modules, special lifting assemblies, etc.

Navy recovery and monitoring ships were positioned along the ground track of the spacecraft. Located 1,000 miles downrange, a range instrumentation ship monitored the spacecraft telemetry during the launch and orbital insertion phases. Radars, antennas and cameras on various military aircraft, ships and ground stations around the world followed the orbiting spacecraft as it sped across their line of sight, handing off the telemetry monitoring and voice relay functions as necessary. If required, mobile stations relocated during subsequent orbits to provide optimum positioning in case of a mission abort.

After one and one-half (or more) orbits of Earth, the Saturn S-IVB third stage rocket was fired over the southwestern Pacific Ocean to propel the spacecraft on its way to the Moon. Somewhere below, an ARIA aircraft would be flying at 40,000 feet recording telemetry data from the spacecraft and relaying voice communications between the astronauts and NASA's Mission Control Center. Below that, at 10,000 feet, several ARRS rescue aircraft orbited at their contingency recovery areas. On the surface of the ocean was a range tracking ship, ensuring capture and relay of key telemetry signals. Slightly further downrange, the primary and secondary recovery ships were on-station in their TLI abort areas in case a problem arose.

Once the spacecraft entered a lunar orbit trajectory, many DoD assets were released since the only successful re-entry into the Earth's atmosphere was along the Mid-Pacific Line and the Atlantic Ocean Line. At this time, the primary recovery force repositioned itself from the mission abort area to the designated primary and secondary landing areas, conducting training exercises until the day before the actual recovery.

During the re-entry phase, several ARIA aircraft orbited along the predicted flight corridor to relay communications to the NASA Mission Control Center just before the blackout period and during the recovery. One Airborne Lightweight Optical Tracking System aircraft was stationed at the point of re-entry to film the Command Module during its fireball stage. ARRS aircraft were stationed uprange and downrange of the Navy's primary recovery force in case the CM fell short or over-shot the End-Of-Mission target point. Once the spacecraft had deployed its main parachutes, and VHF radio contact established, the airspace controller from the Primary Recovery Ship (PRS) handled communications until splashdown and the UDT team took over.

The primary objectives for the recovery force included safely retrieving the astronauts and the Command Module with its equipment and lunar soil samples. Once these objectives were met, they would also try to recover the parachutes and apex cover. During the Apollo program, the Primary Recovery Ship was always an aircraft carrier (CVS) or amphibious assault ship (LPH), with one or two support vessels such as a refueling tanker and a destroyer. NASA preferred using aircraft carriers since their ability to launch and land fixed wing aircraft provided more options for handling logistics and emergencies. For several missions, a communications relay ship was present as well, primarily to handle the overflow traffic created by media reporters covering the event.

Fig 3F—To retrieve Apollo astronauts from the sea, the Navy decided to replace the horse collar sling with a safer "Billy Pugh" rescue net. This photo shows Jim Lovell being hoisted up to a helicopter while undergoing water egress and recovery training in the Gulf of Mexico in October 1968.

Embarked on the PRS were Navy Underwater Demolition Team (UDT) swim teams, a helicopter squadron equipped with search and rescue aircraft, early warning radar aircraft, and logistics support aircraft. Well before the splashdown event, the primary recovery forces underwent extensive training to ensure effective communications and cooperation within and between all units. When the quarantine was in effect, the PRS also carried two Mobile Quarantine Facilities (MQFs), a primary and a backup, as well as other biological containment garments for the astronauts.

The post-flight landing and recovery procedures were carefully thought out by joint NASA/DoD planners, and thoroughly documented in NASA publications such as the "Apollo Operational Recovery Procedures Manual (MSC-01856)." There were serious hazards still to be faced, both natural and man-made. The weather and sea state were very important, since large swells could make the process of attaching the sea anchor, flotation collar and rafts to the CM very difficult. UDT personnel had to work very cautiously around the Command Module since it had unexpended, and highly toxic, hypergolic fuel from its maneuvering thrusters. The CM might also have unfired pyrotechnics (parachute mortars, etc.) that could cause serious harm to anyone near them when they went off.

Regardless of what the manuals said, the vast surface of the Pacific Ocean was not a controlled environment. Curious sharks made appearances while training exercises were underway, especially if the rehearsal was held right after lunch when fresh garbage chummed the waters behind the ship. They once interfered with an actual recovery by appearing as the USS *Yorktown* (CVS-10) was approaching the *Apollo 8* spacecraft for retrieval. The CM's huge parachutes with their long lines still floating in the water could easily ensnare an unwary swimmer. During the recovery of *Apollo 9*, the down-wash from the helicopter rotor blades flipped over one of the life rafts. By doing a number of practice recoveries before the real event, the team was able to confront and plan for these types of unexpected situations.

Following the completion of a recovery operation, NASA's special equipment such as lifting hooks, flotation collars, boilerplates, rafts, radios, etc., were cleaned, returned to their proper staging area and prepared for the next mission. Post-flight reports and debriefing sessions helped ensure the process was constantly improved throughout the Apollo program.

Technical Note:
MSFN Mobile Facilities

NASA relied on two principal communications networks to support the Apollo program. The Manned Space Flight Network (MSFN) interacted with manned spacecraft in Earth orbit. The Deep Space Network (DSN) interacted with manned craft more than 10,000 miles from Earth, in addition to its primary mission of data collection from deep space probes.

To handle the increased Apollo spacecraft telemetry monitoring requirements, NASA needed to expand the MSFN beyond those capabilities provided by existing land-based tracking stations. In conjunction with the DoD, they developed two types of mobile systems that would cover mission critical areas when required.

The first type of mobile facility was the sophisticated Apollo Instrumentation Ship (AIS). In 1966, three World War II era oil tankers ships (T2 hulls) were converted into this new class, becoming the USNS *Vanguard* (T-AGM 19), USNS *Redstone* (T-AGM 20) and USNS *Mercury* (T-AGM 21).

A seventy-two-foot section was added into the middle of the ship, housing an extensive array of electronic equipment, and four large antennas (sometimes covered with white radomes), were placed on the deck. These systems were able to interact with a spacecraft for about ten minutes during each orbit.

For Apollo lunar flights, the AIS primary mission focused on the critical "orbit

Fig 3G—Four large antennas on the USNS *Vanguard's* deck captured general purpose telemetry, C-band radar, satellite data link, and the Unified S-band system used by Apollo to carry the functions of tracking, voice, TV, telemetry and command across a single carrier wave.

Fig 3H—The nerve center of USNS *Mercury* showing some of its sophisticated electronic equipment. Given its position in the Pacific, this ship would occasionally carry an astronaut who would perform the Capsule Communications (CAPCOM) function as the spacecraft flew overhead.

injection" phases—initially for Earth orbit injection (*Vanguard*) and then lunar orbit injection (*Redstone* and *Mercury*). These seagoing tracking stations were placed under the command of the Military Sea Transportation Service (MSTS) but their "range rat" crews were mostly contractor personnel from electronics companies such as Bendix, RCA and ITT.

During the critical TLI burn of the Apollo S-IVB engine, the spacecraft would travel a distance of about 1,800 miles. While two MSTS "injection" ships were positioned to cover the beginning and the end of the TLI burn corridor, they were unable to receive spacecraft telemetry and maintain voice communication with the astronauts during that period. Additionally, while the TLI burn was intended to occur during the second revolution of Earth, it might be delayed for one more. The geographical location of the burn event moved with each orbit, so NASA needed an airborne platform capable of reacting within the ninety-minute orbital period.

A somewhat similar situation could occur during re-entry. Weather, sea conditions, or other factors could cause a change in the recovery area. During re-entry, NASA needed to have a voice relay capability provided to the CAPCOM before and after the communications blackout with the spacecraft. It also needed HF-based homing to determine the spacecraft's location and supply voice relay from the spacecraft to recovery forces. The name of the game was mobility.

To fill these gaps, a new concept in tracking stations was developed—a high-speed aircraft containing the necessary instrumentation to assure spacecraft acquisition, tracking and telemetry data recording.

NASA and the USAF jointly developed a solution called the Apollo Range Instrumentation Aircraft (ARIA). Essentially, eight Boeing C-135A jet transport/cargo aircraft (the military version of the Boeing 707) were modified in 1967 to contain the necessary elements. Redesignated as EC-135N, this highly mobile platform had the ability to receive and transmit astronaut voice communications, and record telemetry information from the Apollo spacecraft. Using a seven-and-a-half-foot steerable parabolic antenna mounted in a radome on its nose, ARIA received telemetry signals via VHF and S-band systems.

Fig 3I—The USAF added a film pod and recording equipment to four of the EC-135 ARIA aircraft so they could photograph a target in flight. The white teardrop-shaped pod can clearly be seen attached to the fuselage just in front of the left wing of aircraft #10327. ALOTS filmed the orbital insertion of *Apollo 11* during the launch and its atmospheric re-entry during the return.

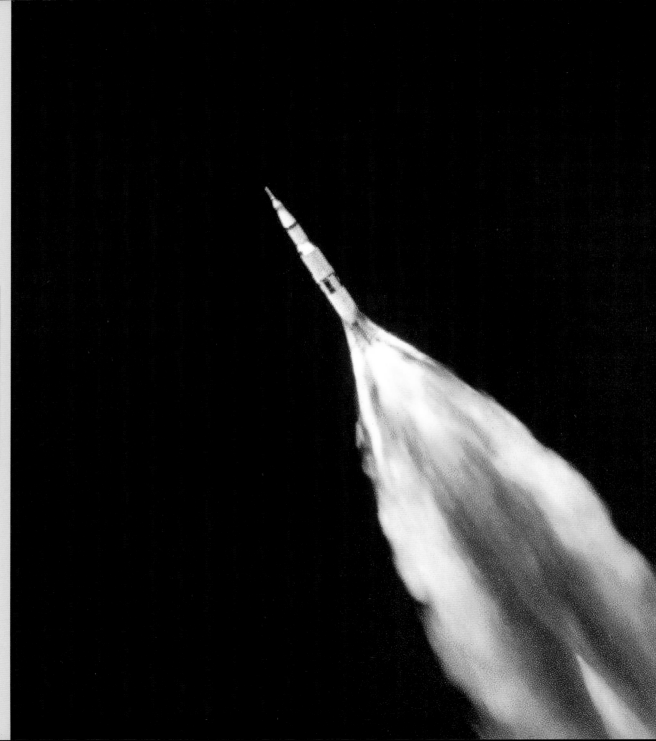

Fig 3J—An ALOTS aircraft was assigned to film Apollo launches just before the Earth-orbit injection event. On July 16, 1969 this photo was taken from 35,000 feet of *Apollo 11* showing the five F-1 first stage engines still producing thrust. Soon thereafter, these booster rockets fell away and five J-2 second stage engines propelled it into Earth orbit.

Vital information ranging from the position of switches on the spacecraft console to heartbeat rates was recorded and retransmitted to ground stations. These droop-nosed aircraft became operational in January 1968 and operated worldwide during the Apollo program.

AFETR was selected to operate and maintain ARIA using Patrick Air Force Base as the home facility. McDonnell-Douglas and Bendix Corporation were the contractors for the design, aircraft modification and testing of the electronic equipment.

When needed, ARIA could be easily integrated into the worldwide manned space flight communications network, providing two-way voice relay between the Apollo spacecraft and the Mission Control Center at Houston, Texas.

Four of the EC-135N aircraft could be reconfigured to carry an Airborne Lightweight Optical Tracking System (ALOTS). This included an external camera pod attached to a special cargo door, fitted on the left side in front of the wing. This configuration was designed to film the spacecraft at high altitude during launch or as it entered the Earth's atmosphere (i.e., during the fireball period) during the end of mission.

When prepared for launch and re-entry missions, a 200-inch lens was installed on the ALOTS camera. The pod was loaded with 1,000 feet of 70mm still film that recorded at forty frames per second. Filming began when the spacecraft was at an altitude of 400,000 feet and continued until the film was exhausted roughly 100 seconds later.

Several ARIA were positioned along the 1,500 mile long re-entry corridor to establish voice and data links with the astronauts just prior to atmospheric reentry and relay this information back to NASA's Mission Control Center (MCC). During the re-entry event, very high temperatures from the Command Module's ablative heat shield created substantial ionization around the spacecraft, so radio signals couldn't get through. One ALOTS was assigned to film the CM and its associated components during this blackout period so if anything went wrong, NASA would have some evidence of what occurred.

Once the CM re-entry occurred, NASA had no control over what happened and they needed a way to monitor the ongoing activities and status of the flight crew. Generally, as soon as the Navy recovery forces established a UHF radio link with the Command Module after the blackout ended, the ARIA would monitor that frequency and pass it back to the MCC. While the only people talking on the circuit included the ship, the recovery aircraft and the astronauts, flight controllers in Houston wanted real-time information about the recovery process.

Both the AIS injection ships and the ARIA instrumentation aircraft were used for general USAF missile range support duties when not required for Apollo missions.

Fig 4A—HS-4 Helicopter #67 returns to the USS *Princeton* after a practice recovery of *Apollo 10*.

Chapter Four
Initial Apollo Fights

The Apollo program officially started with a number of unmanned missions in 1966 designed to test various rocket, spacecraft and operational support systems. The first launch was the unmanned Apollo-Saturn 201 (AS-201) in February. Though only a thirty-seven minute suborbital flight, it validated a number of key design and operations systems. In July, AS-203 was placed into Earth orbit so additional launch vehicle and spacecraft subsystems could be tested. In late August, AS-202 rocketed off the launch pad at Kennedy Space Center to verify adequacy of the heat shield for re-entry at high speeds. The spacecraft, CM-011, landed in the Pacific Ocean 400 miles east of Wake Island and was recovered by the USS *Hornet* (CVS-12).

NASA was on track for an initial manned launch planned for February 1967. However, on January 27, 1967, the *Apollo 1* crew of Gus Grissom, Ed White and Roger Chaffee were doing a full launch rehearsal in their Command Module on top of an unfueled Saturn 1B rocket. A fire broke out inside the spacecraft, probably caused by a spark from a wiring issue. Due to the 100% oxygen environment of the Block-I CM, the fire rapidly engulfed its interior. Unfortunately, the hatch opened inward and a number of latches had to be manipulated with a ratchet while the cabin pressure was reduced to sea level, a process that took ninety seconds under ideal conditions. By the time technicians were able to get to the spacecraft on the launch tower, all three crew members had perished from smoke inhalation and fire.

The manned flight program was delayed more than a year as a number of safety improvements were made, many of which related to the Command Module. The new Block-II CM had a normal air environment plus an access/egress hatch that swung outward and could be quickly opened from within.

However, NASA was able to make progress with its launch vehicle testing by continuing the unmanned flight program. In November 1968, *Apollo 4* provided the first full test of the three-stage Saturn V rocket. It carried an Apollo Command and Service Module

(CSM) into Earth orbit and was considered a major success. In January 1968, *Apollo 5* was the first test flight of the Lunar Module (LM). NASA and its contractors verified the correct operation of ascent and descent stages, propulsion systems, and engine restart capabilities, and evaluated the overall spacecraft structure including the LM. The April 1968 flight of *Apollo 6* provided a final qualification of the Saturn V launch vehicle and Apollo spacecraft for manned missions. In addition to verifying the proper functioning of those systems, the flight also validated the operational readiness of mission support facilities and the DoD spacecraft recovery program.

Apollo 7, the first manned Apollo mission, lifted off from Kennedy Space Center in a blaze of orange-colored flame in October 1968. With its crew of Wally Schirra, Walt Cunningham and Donn Eisele, the spacecraft achieved an elliptical Earth orbit about ten minutes after launch. Over the course of their ten-day, 163-orbit mission, the CSM propulsion system worked perfectly, a major factor in allowing later flights to enter and exit lunar orbit. On October 22, the spacecraft splashed down in the Atlantic Ocean southeast of Bermuda, just over a mile from the planned impact point. The CM turned Stable 2 (upside down) immediately after impacting the water, so the ELS righting bag system had its first try-out. They worked fine, flipping it right side up in just a few minutes. The crew was picked up by helicopter and landed on the deck of the USS *Essex* (CVS-9) at 0820 local time. The spacecraft was hoisted aboard by the ship's B&A crane roughly forty-five minutes later.

The First Lunar Mission

In December 1968, NASA launched the first manned flight to the Moon, using the mighty Saturn V rocket. The *Apollo 8* astronauts, Frank Borman, Jim Lovell and Bill Anders, became the first humans to leave Earth's gravity, the first to orbit another celestial body and the first to see the far side of the Moon. *Apollo 8* completed ten orbits of the Moon before heading back to Earth. While in lunar orbit they made a Christmas Eve television broadcast watched by millions of Earthlings and saw the first Earthrise from the Moon. The success of this historic mission paved the way for the completion of a lunar landing before the end of the 1960s.

John Stonesifer recalls: While the USS *Yorktown* was en route to the *Apollo 8* recovery area, Twentieth Century Fox was filming a segment of the movie *Tora, Tora, Tora* onboard. Thirty propeller-driven aircraft, mostly former U.S. Navy training planes painted to resemble Japanese warplanes, were launched off the flight deck. Shortly after we left San Diego for Pearl Harbor, these planes took off to simulate the attack on Hawaii in 1941 that brought the U.S. into World War II.

Fig 4B (opposite)—This view of the rising Earth greeted the *Apollo 8* astronauts when they came from behind the Moon after the lunar orbit insertion burn. They became the first humans to ever view the far side of the Moon.

Fig 4C (below)—This spectacular photograph was taken by a USAF ALOTS aircraft flying at 40,000 feet. It shows the *Apollo 8* spacecraft entering the Earth's atmosphere at around 400,000 feet over the South Pacific on December 27, 1968.

The atmospheric re-entry of *Apollo 8* was filmed by an ALOTS photo aircraft, which captured the fast-moving fireball in 70mm still image format. The recovery forces were led by the USS *Yorktown* (CVS-10). Splashdown occurred at 0452 local time on the morning of December 27, the first landing in darkness for the entire space program. The End-Of-Mission (EOM) point was about 1,000 miles southwest of Hawaii and only two-and-one-half miles from the Primary Recovery Ship. A significant portion of the 1,650-man ship's crew watched the spectacular re-entry from the flight deck and island superstructure. After landing, the CM's parachutes dragged it into an upside-down position (Stable 2) while ten-foot high swells rocked the spacecraft, making Borman seasick.

The recovery team waited until the sky lightened before beginning the recovery phase at 0535, about one half hour after the spacecraft was turned apex-up (Stable 1) by its righting bags. An HS-4 helicopter deployed UDT-12 into position to begin the astronaut retrieval process. The UDT team, composed of LTJG Dick Flanagan, STG3 Bob Coggin and SFC Don Schwab, worked efficiently in the seventy-eight degree water. When signaled that all was ready, the Commanding Officer of HS-4 and pilot of Helicopter #66, Commander Don Jones hoisted the astronauts into his aircraft quickly and headed back to the ship. The *Apollo 8* crew landed on Yorktown's flight deck ninety minutes after splashdown.

The first humans to travel around the Moon were welcomed aboard by Captain John Fifield and NASA recovery team leader John Stonesifer during a brief flight deck ceremony. The astronauts entered the ship's medical spaces to undergo various tests by the seventeen man NASA team, led by Dr. Clarence Jernigan. This four-hour process was interrupted by a telephone call from President Lyndon Johnson. By 0730, the CM had been retrieved and the recovery force departed the mid-Pacific scene. After a formal dinner that evening, the astronauts met with some of the ship's crew in the hangar deck. Navy Captain Jim Lovell presided over the re-enlistment of a few enthusiastic sailors. Following this was the traditional cake cutting party, not a small task considering it involved a seven-foot long, three-foot wide cake that weighed 540 pounds. After a good night's sleep the astronauts boarded a VR-30 squadron C-1A Trader aircraft and catapulted off the ship for a 300-mile flight to Hawaii.

This was the inaugural mission for several DoD enhancements to the recovery process. The major communications relay ship USS *Arlington* (AGMR-2) was assigned to the recovery force to assist with getting the media's reports sent back to the mainland in a timely manner. In addition to the Apollo Range Instrumentation Aircraft (ARIA) monitoring activities, an ALOTS aircraft photographed the re-entry burn period. *Yorktown* had also been fitted with a new military Tactical Communications Satellite (TACSAT) system giving it the ability to back up its normal HF radio links. Several hours after recovering the Command Module, an ARRS HC-130H performed a Surface-to-Air-Recovery (STAR) pickup over the flight deck using the Fulton Skyhook. The aircraft snagged a line held aloft by a balloon to retrieve a container with medical samples and pool news film, then flew back to Hawaii.

Fig 4D—A UDT swimmer snaps a steadying line to the sea anchor ring in the front of the *Apollo 8* CM. The recovery loop is above the hatch (adjacent to the upper frogman's flipper) and must be facing the ship before the B&A crane can lift the spacecraft out of the water, since it hangs at a thirty degree angle when being lifted. During the hoist operation, two steadying lines are used by the ship's crew to keep the CM from swaying and to guide it onto its transport dolly on the elevator.

One Final Earth Test

In March 1969, the second manned launch of a Saturn V rocket lifted *Apollo 9* into low Earth orbit for a ten-day mission. This was the first manned flight of all three of the primary Apollo spacecraft modules. Its three-man crew, Jim McDivitt, Dave Scott and Rusty Schweickart, tested several aspects critical to landing on the Moon including the Lunar Module engines, backpack life support systems, navigation systems and docking maneuvers. They performed the first manned flight of a LM, the first docking and extraction of a LM, a two-man spacewalk that checked out the new Apollo spacesuit, and the first docking of two manned spacecraft. The splashdown on March 13 was 180 miles due east of Bahamas and occurred within sight of the USS *Guadalcanal* (LPH-7).

The Second Lunar Mission

In May 1969, NASA launched the *Apollo 10* mission as a full dress rehearsal for a Moon landing. After lunar orbit had been achieved, astronauts Gene Cernan and Tom Stafford disengaged the Lunar Module from the Command and Service Module, leaving John Young alone to orbit the Moon several times. The LM descended to roughly nine miles above the lunar surface. Except for the lack of an actual landing, the mission went exactly as a lunar landing would have gone, both in space and on the Earth, where NASA and DoD tracking and control networks were thoroughly vetted.

USS *Princeton* (LPH-5) led the recovery force. Additional sophisticated military technology was injected into this recovery operation. In the mid-Pacific, standard maritime radio navigational aids such as Long Range Aid to Navigation (LORAN) were unavailable. To improve *Princeton's* navigation capabilities, a Navy Navigation Satellite System terminal was connected into the ship's regular navigation system. The resulting TRANSIT satellite data greatly enhanced the ship's position determination and course plotting abilities.

On May 26, as dawn was breaking, the Command Module (call-sign *Charlie Brown*) re-entered Earth's atmosphere (400,000 feet altitude) at a velocity of 36,314 feet per second heading almost due east. An ARRS HC-130H stationed 230 miles up-range (to the west) of the Primary Recovery Ship reported the initial visual contact. Soon a number of Navy ship and aircraft personnel watched the traverse of *Charlie Brown* as it sped across the clear, dark sky. It was as bright as a comet, leaving a long trail of fire.

During its parachute descent near the recovery ship, the faint dawn light made it clearly visible to the recovery helicopters, which circled the CM until it splashed into

Fig 4E—The *Photo* helicopter was close enough to the descent path of the *Apollo 10* Command Module to catch this photograph just prior to splashdown.

Commander Chuck Smiley recalls: Just before dawn on May 26, 1969, I lifted Helicopter #66, call-sign *Recovery Three*, off the deck of the USS *Princeton*. We maintained a 5,000-foot high orbit around the ship, counting down the seconds until the spacecraft's re-entry. Suddenly, a speck of light appeared in the sky just exactly where we had been told it would be by the NASA folks. The speck grew a tail of flame and appeared to be climbing, which was an optical illusion. It was actually descending at very high speed through the upper atmosphere. The object continued until it was directly overhead. Then the flame flickered out, which meant it was in a vertical or straight down trajectory!

A thought momentarily flashed through my mind that the spacecraft might actually be directly overhead and plummet right on top of us. Ah, a billion to one chance? Wouldn't that be embarrassing on worldwide TV! Well, I didn't have time to waste on that sort of thought and got back to handling my Sea King. Shortly afterward, the Pacific. It hit the sea at 0552 local time, just one-and-one-half miles from its planned impact point and about three-and-one-half miles from the carrier. The End-Of-Mission location was 2,400 miles south of Hawaii and a few hundred miles east of American Samoa. The winds were light and the water calm with only three-foot waves, so the Command Module remained upright (Stable 1) and the righting bags were not triggered for inflation.

For safety reasons, the recovery team waited a few minutes until the sky was lighter. Commander Chuck Smiley, the HS-4 Executive Officer, piloted Helicopter #66 and was assigned the radio call-sign of *Recovery Three*. In addition to its aircrew, the Sea King carried three UDT swimmers and a NASA flight surgeon.

Recovery Three made its first approach to the CM, dropping QM3 Michael Mallory who cleared away the parachute lines and attached a sea anchor to the spacecraft. On the second helicopter pass, LTJG Wes Chesser, the UDT-11 swim team leader, and BM1 Louis Boisvert, jumped into the water while the aircrewmen shoved the flotation collar and life rafts out of the cargo hatch. After securing the spacecraft, they assisted the astronauts into the hoist raft. One by one, the three astronauts were hoisted up in the Billy Pugh net, helped through the hatch and seated on a canvas troop seat in the cargo compartment. The astronauts exchanged their space flight "constant wear garments" for NASA blue flights suits before being brought back to the carrier. The *Apollo 10* crew descended from the helicopter onto the ship's flight deck less than an hour after splashdown.

After addressing the welcoming group, the astronauts went into the medical area for physical exams and refreshments. The CM was recovered by the PRS one-and-one-half hours after landing. *Princeton* then steamed toward American Samoa for eight hours until it was within the flight range for a Sea King helicopter. The astronauts were flown by three HS-4 helicopters to Pago Pago for transit back to Houston.

Before returning from Pearl Harbor, *Princeton* provided further assistance to NASA by conducting an open ocean verification test of the biological decontamination and quarantine procedures that would add significant complexity to upcoming lunar landing missions.

The *Apollo 8* and *Apollo 10* recoveries, along with the follow-on decontamination testing, were valuable learning experiences in terms of handling lunar mission recovery

I got my first view of the spacecraft's Earth Landing System with the flashing light, three giant parachutes and the Command Module silhouetted against the pre-dawn sky. From the top of the chutes to the bottom of the spacecraft is roughly 210 feet, equivalent to a twenty-story building descending through the sky just a mile away. What a thrilling sight that was.

In keeping with the *Peanuts* comic strip theme of the mission, HS-4 crewmen painted a sign on the underside of the helicopter saying "Hello der Charlie Brown" to greet the space travelers as they made their way up.

Fig 4F—This quintessential Apollo recovery photo was captured by HS-4 photojournalist Milt Putnam just as the sun was rising. Two of the astronauts have exited the CM and the crew of Helicopter #66 is preparing the Billy Pugh net for hoist operations.

Chuck Smiley recalls: "On the transfer to Pago Pago, I asked my copilot Scotty Walker to move to the rear and let Tom Stafford have the copilot's seat. The opportunity to share a cockpit with a guy that just came back from the Moon was too good to miss."

requirements. The NASA Recovery Team, Air Force ARIA, ALOTS and ARRS units, Navy helicopter squadron HS-4 and swim team UDT-11 had all gained confidence in their procedures, equipment and intra-unit communications. New technologies were providing a greater margin of assurance for success. Much of the recovery team was now well prepared to play its role one of the greatest scientific achievements in human history—landing men on the Moon and returning them safely to Earth.

Fig 4G—The *Apollo 10* crew was greeted on the flight deck by NASA recovery team leader Dr. Don Stullken (white shirt) and *Princeton* CO Captain Carl Cruse (just out of the photo on left). The astronauts addressed an enthusiastic crowd of ship's personnel and media.

Fig 5A-A view through the MQF front door into the lounge area.

Chapter Five
Earth Contamination Issues

At the beginning of the Apollo program, while lunar sample-return missions were being planned, protection against back-contamination of Earth was a critical issue. No one knew for sure if there were any pathogens or germs on the Moon that might inadvertently be brought back and cause widespread damage. The Space Science Board of the National Academy of Sciences created a council of scientists, healthcare professionals and government personnel called the Interagency Committee for Back Contamination (ICBC), to explore this issue. They recommended to NASA that anything coming in contact with lunar soil be quarantined from the Earth's biosphere until proven harmless.

NASA eventually decided to quarantine astronauts, equipment and soil samples returning from the lunar surface for a period of twenty-one days. Since no one had any experience with a Moon germ, NASA's quarantine solution was based on the "best practice" for germs on Earth. Twenty-one days was considered the outside incubation period for most harmful bacteria. The quarantine "clock" started from the moment the Lunar Module hatch was closed on the Moon's surface. About three-and-one-half days were consumed during the flight back to Earth. While the astronauts had to spend three-and-one-half more days aboard ship (and aircraft) in transit from the splashdown point to Houston, the lunar samples were flown from the recovery ship to the Lunar Receiving Laboratory (LRL) at Johnson Space Center. During the ensuing two weeks, both astronauts and Moon rocks were extensively tested for the presence of live lunar microorganisms.

Since the initial crew, equipment and sample recovery would be accomplished in the middle of the Pacific Ocean, NASA devised unique solutions for this risky transition period of the quarantine process. These consisted of:

- Mobile Quarantine Facility (MQF) to house the astronauts, soil samples, space suits and other equipment while en route to the LRL
- Biological Isolation Garments (BIG suit) to encase the returning astronauts for their helicopter ride from the floating CM to the recovery ship
- Some liquid disinfecting agents (such as betadine and sodium hypochlorite) to clean potentially infected surfaces.

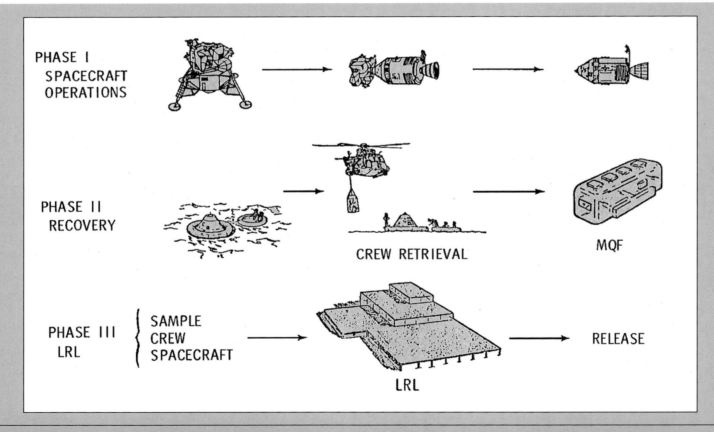

Fig 5B—This NASA graphic provides an overview of the astronaut quarantine process. Phase 1 encompassed the flight back to Earth, Phase 2 was the transitional phase during the sea recovery operation and return to the Lunar Receiving Laboratory where they entered Phase 3, the final fourteen days.

Mobile Quarantine Facility (MQF)

The MQF was essentially a travel trailer, modified to prevent the escape of infectious biological agents. It was designed by Melpar, a subsidiary of American Standard, to transport the lunar explorers and material (rocks, dust, etc) from the splashdown location to the LRL. The MQF was based on a standard 1969 Airstream trailer and used many of the normal travel trailer construction features. However, instead of wheels, the trailer base was a thirty-five foot long metal shipping pallet that provided increased structural strength to the whole unit for crane-lifting and transport purposes. Equipped with special tie-downs, it was designed to be carried in the hangar bay of an aircraft carrier and in the hold of a cargo plane.

The facility was composed of four major areas—crew lounge, galley, bunks and lavatory. It was built to house six people for up to ten days, having sleeping bunks, toilet facilities, kitchen equipment, chairs, table and medical diagnostic equipment. In an Apollo recovery environment, the MQF held five people—three astronauts, one doctor and one MQF technician. All the lunar-contaminated items, other than Moon rocks, were stored inside during the period when they were being transported from the splashdown point to the LRL, roughly three-and-one-half days.

Providing "a bio-home away from home," the MQF maintained a negative pressure differential *vis à vis* the outside air and filtered any air escaping to the outside.

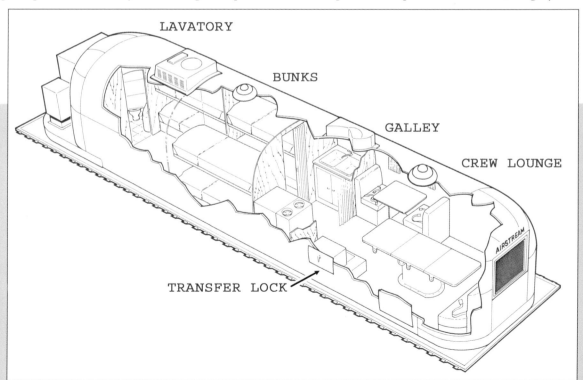

Fig 5C—Melpar released this depiction of the MQF as part of their original press kit in 1969 that shows the four main living areas. The location marked "Transfer Lock" is the device by which material was passed into or out of the MQF so that a biological isolation could be maintained in both directions.

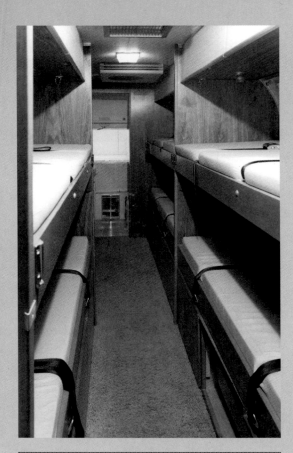

Items passed into or out of the MQF went through a submersible transfer lock to prevent the movement of germs in or out of the trailer. Specially prepared meals were passed into the facility via the transfer lock and heated in a unique (for its time) microwave oven. Two independent heating and air conditioning units maintained a comfortable inside temperature. Basic medical diagnostic equipment for post-flight crew examinations and tests was included as well as a main table that could be quickly converted to an examination stand.

The MQF communications patch panel provided facilities for two-way telephone communications, TV reception, shipboard motion picture audio, and an output for a shipboard public address system. The lounge area contained six swivel-type aircraft seats and a removable multi-purpose table, which was used for medical examination, eating meals and other purposes. The galley contained a refrigerator, sink, microwave oven, and the decontamination transfer lock. The sleeping area was equipped with six bunks, which had individual reading lights and storage space for personal items as well as space suits etc. Aft of the bunks on the left were closets where the astronaut's clothing was kept. The bathroom had a toilet, sink, hot-water heater, bathtub and shower, plus an emergency exit.

Controls for the entire MQF were mounted on a single mode panel, allowing the MQF technician to monitor all environmental and power systems constantly. From here, he

Fig 5D (above)—The bunks came complete with seat belts in case the recovery ship or transport aircraft encountered rough weather. The bins under the lower bunks were used for storing bulky items such as space suits and helmets, while drawers above the upper bunks were for smaller items. Visible on the ceiling is one of the negative pressure fans that ensured containment of the MQF's atmosphere. Only the prototype MQF was carpeted—the rest used linoleum.

Fig 5E (right)—The Control Panel provided the central coordination point for all the electronics in the MQF. From here, the MQF technician could monitor the general state of the facility environment.

could start up the diesel generator, if required. He could also place phone calls, patch through TV signals, handle lighting and air conditioning requirements, etc. Visual and audible signal alarms were provided for a variety of problem situations. A self-contained generator and two rotary converters provided power for the facility. In the event of a power failure, the MQF automatically switched to an emergency backup system. Another contingency system provided oxygen to the occupants should the host aircraft lose its pressurization.

Because it was designed to be biologically sealed aboard the recovery ship, then later flown in a cargo aircraft, the MQF had multiple systems for power, communications and air conditioning. For instance, the electrical system was capable of interfacing with ships, aircraft and transportation vehicles as well as using its own diesel generators. The outer shell of the all-aluminum trailer also had "blowout panels" to mitigate against the risk of a break in the quarantine if a sudden decompression in the cargo hold of the aircraft occurred during flight.

The quarantine plan called for two other NASA professionals to be inside the MQF with the astronauts after an Apollo recovery. One was the MQF technician who managed the operation of the trailer itself and would troubleshoot any problems that might arise. He also assisted with unloading the Moon rocks and soil samples from the Command Module and repackaging them for transport via airplane to the Lunar Receiving Lab in Houston. The other was a NASA flight surgeon (doctor) who took care of the astronauts and ran the myriad of medical tests required of all astronauts after a flight.

Once the Command Module had been recovered and secured near the MQF in the ship's hangar bay, a plastic-sheet tunnel connected them together. This allowed Moon rocks, film, space suits, and other equipment to be transferred into the MQF while maintaining the biological containment. The astronauts were free to go back inside the CM to retrieve personal belongings as well.

Fig 5F—Another drawing released by Melpar depicted how the CM would be connected to the MQF via a plastic tunnel. This created a biologically controlled connection between the two components, allowing the astronauts and the MQF technician to off-load items from the CM without breaking the quarantine. This drawing reflects the initial CM egress plan, which called for the astronauts to stay inside the CM as it was hoisted onboard the recovery ship and connected to the MQF.

To most humans, the MQF seems like a very cramped place for five people to stay over a three-day period. But *Apollo 14* astronaut **Edgar Mitchell recalls,** "For three astronauts who had just spent eight days weightlessly bumping around each other in the Apollo CM, with no privacy (nor gravity) for eating, sleeping or personal hygiene activities, it seemed like a palace."

Once inside their special "biosphere," the astronauts were allowed to leave only after the MQF was backed up to the receiving dock at the LRL of Johnson Space Center and mated to an air-tight hatch. They went directly into another biologically isolated environment where they stayed for the rest of the quarantine period.

NASA eventually ordered four MQF's from Melpar. They were all built by Airstream Corporation at their Jackson Center, Ohio facility. Two MQFs were assigned to every quarantined Apollo recovery, a primary one for the astronauts and a backup in case the first one encountered a problem or there was contamination of members of the recovery team.

Biological Isolation Garment

During the early Apollo planning stages, it was assumed the astronauts would remain inside the Command Module while being hoisted aboard the recovery ship. They would emerge directly into the plastic tunnel that connected the MQF with the CM, ensuring the necessary biological isolation requirements. However, for various reasons, this was changed just before the initial Apollo missions.

NASA then had to devise a method of transporting the astronauts back to the MQF aboard the ship while the spacecraft was still floating in the ocean. NASA assumed the CM's internal atmosphere would not be contaminated since the spacecraft's environmental control system should have filtered out airborne bacteria during the long return trip from the Moon. Thus only the astronauts—not the atmosphere of the Command Module—were likely to harbor biological contaminants or pathogens. To ensure they would not infect the world on their short trip from the spacecraft to the MQF, NASA's plan called for the astronauts to be "bagged" in a Biological Isolation Garment (BIG suit). A Navy UDT diver, termed the "decontamination swimmer," was designated to open the spacecraft hatch after its return to Earth. Clad in a BIG suit himself, he would toss these garments into the spacecraft and then immediately slam the hatch shut. Only after they had donned their own BIG suits, zipped up head to toe, would the astronauts exit the spacecraft into the recovery raft. The BIG suits would be removed after they were sealed in the MQF aboard ship.

The BIG suits were fabricated from a single piece of lightweight nylon cloth, which completely covered the wearer and served as a biological barrier from the Earth's biosphere. A visor/mask assembly provided limited visibility to the wearer. The garments came in three sizes—small, medium, and large—with adjustment straps in key places.

There were two types of BIG suits. Type A was designed to be worn by potentially contaminated individuals, i.e., the astronauts. The face mask was equipped with an air-inlet flapper valve and high-efficiency air outlet filter, designed to filter out contaminants the wearer might exhale.

Type B was for use by personnel who were physically exposed to the astronauts or contaminated surfaces and equipment. This face mask came with an air-outlet flapper valve and air-inlet filter, designed to filter possible contaminates from air the wearer might breath in. The initial plan called for the suiting up the UDT "decon" swimmer and recovery helicopter crew, but this was scaled back to just the UDT swimmer. It was felt the garment might interfere with the aircrewmen's activities. Additionally, if the UDT swimmer thoroughly scrubbed the astronauts before they were hoisted up from the raft, there would be no Moon germs on the outside of their BIG suits anyway.

The BIG suits were only used for one recovery—*Apollo 11*. The astronauts, especially Michael Collins, found them extremely uncomfortable to wear in the tropic heat and humidity while bobbing in the decontamination raft during the recovery process. For *Apollo 12* and *14*, the only other ones subject to quarantine, NASA switched to using flight suits, respirators and special sneakers.

Biological Decontamination Process

The UDT "decon" swimmer was also provided with some "alien germ" cleansing agents—betadine and sodium hypochlorite. Betadine is well known in the medical field as a skin cleaner or disinfectant hand wash and is used for cleansing hands prior to surgery

Fig 5G—UDT decontamination swimmer Lieutenant Clancy Hatleberg is shown wearing a Biological Isolation Garment aboard NASA's M/V *Retriever* prior to astronaut egress training in May, 1969. The suit was not difficult to don while in an open environment, but presented interesting challenges to the astronauts when they were confined inside the bobbing CM on the ocean surface.

and other aseptic procedures. Sodium hypochlorite solution, commonly known as bleach, is frequently used as a disinfectant and as a bleaching agent.

The UDT decon swimmer, wearing a BIG suit, would assist each astronaut out of the Command Module hatch and into the work raft. After all three astronauts had exited, the CM hatch was closed. The decon swimmer used a special glove to wipe the exterior vents and hatch of the Command Module with betadine to ensure no pathogens had escaped via those openings while they had been venting into the Earth's atmosphere.

The possibility of Moon dust lingering in the CM and getting on the outside of the BIGs had to be dealt with. With the astronauts in the recovery raft, they and the decon swimmer scrubbed each other down with sodium hypochlorite. Betadine was not used for this since it was a wetting solution that would make the BIG suits permeable, reducing the quarantine effectiveness.

After all the scrubbing was complete, and the astronauts had been hoisted out of the decontamination raft, the final responsibility of the decon swimmer was to place all of his equipment into that raft and sink it.

Quarantine Training

Part of the NASA culture was to practice as many of the mission activities as possible and prepare for various contingencies. The prototype MQF was loaded aboard several ships to verify compliance with hangar bay tie-downs, power and communications system interfaces, etc. It was also loaded into the cargo hold of a USAF C-141 to ensure all the systems, including the MQF's pallet base, would operate in that environment. The only simulation not run involved a rapid decompression scenario to test the operations of the MQF "blowout panels" during an in-flight emergency. The trailer was also taken to Johnson

Fig 5H—The MQF is being prepared for insertion into the cargo hold of a USAF C-141 Starlifter. For minor movement, such as around the hangar bay of an aircraft carrier, it could be fitted with casters and towed. For longer range mobility on land, however, NASA's plan called for carrying it on the bed of a wheeled transport.

Space Center for evaluation of the plastic tunnel that connected the Command Module with the MQF to maintain the at-sea biological isolation of the lunar-exposed equipment and soil samples.

Lieutenant Clancy Hatleberg of UDT-11 was undergoing training exercises in San Diego related to the *Apollo 10* recovery when he was selected to be the decontamination swimmer for *Apollo 11*. As his team embarked aboard the USS *Princeton*, Hatleberg instead reported to the Johnson Space Center. On May 24, 1969 he and the *Apollo 11* astronauts sailed a few miles out into the Gulf of Mexico aboard the NASA boat M/V *Retriever*.

For the first and only time before the actual mission, the group practiced going through the quarantine recovery procedures, from donning the BIG suits in the CM, to decontaminating each other in the attached raft and being hoisted up into the recovery helicopter.

Fig 5I—Two months before the real thing, UDT-11 decontamination swimmer Lieutenant Hatleberg, wearing his BIG suit, assists Neil Armstrong out of the CM during recovery and decontamination training. A NASA diver floats in the water nearby in case of an emergency. The boilerplate was lowered into the water by the training ship, hence the righting bags are merely for simulation purposes as the CM was never in a "Stable 2" or upside down position.

Hatleberg remembers: "The water egress training boilerplate BP-1102 was lowered into the water and the astronauts were transferred into it. Then I came along side with the decontamination gear, got set up and helped them get out. We went through the decontamination procedure and the helicopter extraction process."

Technical Note:
MQF R&D Program

Recollections about the development of the prototype MQF from John Blossom the Program Manager for Melpar, a subsidiary of the American-Standard Company. **John remembers:**

No one had built a portable biological isolation chamber like this before, so I spoke with a lot of people in the health field and the military about what kinds of capabilities would be needed. Working with NASA, we came up with a set of requirements that allowed us to build a prototype. I researched a lot of options—we needed something fairly strong but yet lightweight and we didn't have a lot of time to produce the finished product. In the end, the Airstream solution was the best one since we could take advantage of many of the standard construction features for their travel trailer.

During this time, there were still many unanswered questions. For instance, at one point NASA thought they might need to have one on every Navy ship assigned to a lunar recovery mission. In the end, everyone settled on just having the MQF onboard the Primary Recovery Ship and that reduced the number to be built by a great deal.

Airstream assigned Stan Taylor to be the MQF production manager and we got along really well. We built the prototype at their Jackson Center, Ohio plant from November 1967 through February 1968. I designed a heavy-duty frame, which interfaced directly to the USAF 463-L cargo handling system, as the base upon which a thirty-two-foot travel trailer shell was installed. In addition to well-known attributes of the MQF itself, i.e., the redundant power and advanced communications

Fig 5J—Melpar's Program Manager for creating the MQF, John Blossom (closest to camera), discusses the upcoming operational readiness test program for the nearly-completed MQF prototype with Rod Bass of NASA.

capabilities, we also created special support systems such as a single hoisting sling with spreader bars that could handle its 10,000-pound weight.

We finished the prototype MQF on February 29, 1968. Working with NASA, the Navy and the Air Force, we immediately performed various operational readiness tests to ensure it would work in the space mission recovery environment.

For the initial sea trial, it was loaded aboard the USS *Guadalcanal* (LPH-7) in Norfolk, Virginia. We learned a lot about hoisting the aluminum-shelled trailer by crane, such as needing to have a single lift point but using spreader bars so the trailer frame wouldn't buckle nor have external components get damaged. We also learned the best method for chaining it down in the aircraft carrier hangar bay for rough seas and how to ensure full access to the transfer lock. The lock was a very important part of the

Fig 5K (above)—Taken November 29, 1967 in Airstream's MQF manufacturing facility, this photograph shows workers discussing the best process for completing the aluminum covering. Note the one-piece metal base (pallet) upon which the MQF is built.

Fig 5L (left)—The prototype MQF was secured in the hangar bay of the USS *Guadalcanal* for systems evaluation. As this photo was taken, they were testing the submersible transfer lock mechanism.

quarantine operation. Small doors opened into a chamber, about the size of a dishwasher, in which the object to be sanitized was placed. It was then flooded with a bleach-like solution so no germs could survive. In this way, the packages of Moon rocks and lunar film could be safely transferred out while specially prepared meals were transferred in and heated up in the microwave oven.

The next day it was lashed to the deck of the destroyer USS *William M. Wood* (DD-715) for another operational readiness evaluation. An early DoD plan considered stationing destroyers along each of the possible recovery lines in the Atlantic and the Pacific, but that was deemed not to be necessary. Although we proved it could be done, NASA chose not to equip the secondary recovery ships with any quarantine facilities. It was much more economical to position ARRS HC-130H's with pararescuemen in locations where a landing was not probable. No MQF ever went aboard a destroyer again, which is a blessing since we occupants were pitched about quite a bit on this trial run.

The well-traveled prototype was also loaded aboard a USAF C-141 cargo plane at Ellington AFB. It was attached to the cargo pallet transport system and flown around for a while. NASA was extremely thorough in trying to create a viable and effective quarantine system. We also rehearsed the actual recovery process at Johnson Space Center, connecting the MQF to an Apollo boilerplate by a plastic tunnel.

When all was said and done, we found a few problems with the prototype, such as a carpeted floor that might collect lunar dust. We made the required modifications, including switching to a linoleum floor, and built three more MQFs.

The first use of an MQF was on the *Apollo 11* mission for Armstrong, Aldrin and Collins aboard the USS *Hornet*. Four months later, the *Apollo 12* crew of Bean, Conrad and Gordon were quarantined in an MQF, also aboard the USS *Hornet*. The third and final use was in 1971, when an MQF housed *Apollo 14* astronauts Shepard, Roosa

Fig 5M—The MQF was chained to the deck of the destroyer USS *William M. Wood* during a sea trial. The NASA team enjoyed watching the ship's crew taking target practice with .45 caliber automatic pistols on the fantail.

and Mitchell aboard the USS *New Orleans* (LPH-11). The fourth MQF was never used since *Apollo 13* never landed on the Moon due to the in-flight explosion. After the *Apollo 14* mission, NASA decided the Moon harbored no life forms that might be dangerous to the Earth's biosphere and the quarantine process was eliminated.

Fig 5N—This view clearly shows how Melpar specifically designed the MQF pallet, or metal frame, to interface with the USAF C-141 cargo handling system. The trailer was also able to plug directly into the internal power and communications systems on the aircraft.

Fig 6A—The USS *Hornet* compiled an impressive record in World War II, having been involved in the major campaigns of 1944 and 1945. She also conducted three combat tours of duty during the Vietnam War. Shown here in the mid-1960s, she steams into the wind, preparing to launch S-2F Tracker anti-submarine and maritime interdiction aircraft.

Fig 6B—The USS *Hornet* steaming at high speed to her assigned *Apollo 11* station.

Chapter Six
Primary Recovery Force: Ships

During the launch and Earth orbit phase of the *Apollo 11* flight, naval forces were stationed in both the Atlantic and the Pacific. The Atlantic abort recovery team consisted of the mine countermeasure ship USS *Ozark* (MCS-2), the destroyer USS *New* (DD-818), and the fleet tug USS *Salinan* (ATF-161). The Primary Landing Area force in the Pacific included the anti-submarine warfare carrier USS *Hornet* (CVS-12) designated as Primary Recovery Ship. The USS *Goldsborough* (DDG-20) was designated as Secondary Recovery Ship, while the USS *Hassayampa* (AO-145) provided at-sea logistics support. At the last minute, the USS *Arlington* (AGMR-2) was assigned to assist with various communications tasks plus support of the President's visit. Although not a member of TF-130 for this mission, the USS *Carpenter* (DD-825) was stationed halfway between Johnston Island and *Arlington* in support of President Nixon's overwater flight.

USS *Hornet* (CVS-12)

The twelfth U.S. aircraft carrier, or CV-12, was the eighth ship of the American Navy to bear the name *Hornet*. Commissioned in November 1943, she was the fourth of the famous Essex class of aircraft carriers that wreaked havoc on the enemy during World War II. American carriers of that era had a straight (axial) flight deck made of teak for launching and landing aircraft with an island superstructure in the center of the ship on the starboard (right) side. Underneath the flight deck was the hangar deck, divided into three hangar bays, with an armored floor to stop enemy bombs from penetrating into the rest of the ship. The normal aircraft complement of these "fast carriers" during WWII included fighters, dive bombers and torpedo bombers—about 100 aircraft in total.

In February 1944, *Hornet* steamed into the Pacific and spent the next fifteen months in the forefront of combat operations. She ranged throughout the western Pacific with her planes supporting the invasions of the Marshall Islands, the Philippine Islands, the Mariana Islands, Iwo Jima and Okinawa. Overall, *Hornet* participated in nine major campaigns during World War II, established the record for enemy aircraft destroyed, and was awarded a Presidential Unit Citation. Among sailors, *Hornet* was known as a "lucky" ship—attacked fifty-nine times, she was never seriously damaged.

In the 1950's, she underwent several major modernization programs to keep pace with dramatic changes in the world's navies. In 1952, she was upgraded to be able to launch and land jet aircraft and was re-designated an "attack carrier" or CVA. In 1956, an angled landing deck was added to the back half of the ship to allow better aircraft handling flexibility and her bow was enclosed to improve seaworthiness. In 1958, she was converted from an attack carrier to an anti-submarine warfare carrier, or CVS, and refocused on countering the Soviet ballistic missile submarine threat. She conducted numerous cruises to the Far East prior to the start of the Vietnam War.

During the 1960's, *Hornet* spent three tours of duty on Yankee Station, off the coast of Vietnam. Her operational responsibilities included anti-submarine warfare, maritime interdiction, and search and rescue activities for Navy and Air Force pilots whose aircraft were shot down near the coast or over water. Just a few weeks after returning to Long Beach, California from her final Vietnam combat cruise, she was designated Primary Recovery Ship for the *Apollo 11* mission. *Hornet's* newly appointed Commanding Officer was Captain Carl Seiberlich, a 1943 graduate of the U.S. Merchant Marine Academy at King's Point, New York. A naval aviator, he was equally adept at flying aircraft or commanding a ship and had a reputation as a skilled ship handler.

The venerable WWII aircraft carrier was selected by the Navy to perform this unique recovery for many reasons:

- Accessibility—*Hornet* had finished its third and final Vietnam deployment and was stationed on the West Coast of the U.S. mainland at Long Beach. She was scheduled for decommissioning the following year and therefore was not part of any major naval replenishment or refurbishment cycle.
- Availability—*Hornet* was assigned to conduct classified Anti-Submarine Warfare research (Project UPTIDE) from which it could be re-tasked on short notice for eight weeks at a time to handle Apollo mission recoveries.
- Failure management—*Hornet* had a completely redundant propulsion plant, so there was no single point of failure that could cause an aborted recovery mission.
- Agile maneuvering—*Hornet* had four screws and two rudders so it could "nuzzle" up to a small five-ton spacecraft bobbing on the open ocean without risk of smashing it into the depths.
- Familiarity—*Hornet* had recovered the unmanned NASA flight AS-202 in 1966 about 400 miles southeast of Wake Island, so the crew held some familiarity with spacecraft recovery.
- Efficiency—*Hornet* had been rewarded with a battle "E" efficiency rating the previous year, meaning its crew was extremely well prepared to handle any emergency.

- Leadership—Commanding Officer Captain Seiberlich had been Navigator of the USS *Intrepid* (CVS-11) when they developed *Gemini 3* recovery procedures. He was familiar with manned spacecraft recovery techniques and had an existing relationship with NASA officials.

The June selection of *Hornet* as PRS for the mid-July *Apollo 11* space flight was very short notice for this type of operation. Like her predecessor's *Yorktown* and *Princeton*, TF-130 assigned *Hornet* four distinct mission responsibilities—launch abort, TLI-burn abort, deep space abort and End-of-Mission recovery. During the launch and Earth parking orbit phase of the *Apollo 11* mission, *Hornet* would be positioned on the Mid-Pacific launch abort line in case of a launch problem. During the second Earth orbit, when the spacecraft performed its TLI rocket burn, *Hornet* was required to be "on station" in the TLI abort area in case of a malfunction. During the ensuing eight days, TF-130's plan called for *Hornet* to loiter along the Mid-Pacific deep space abort line in case the mission could not fulfill its Moon landing activities. Finally, she had to end up at the End-Of-Mission (EOM) target point, almost 400 miles to the north, so her crew could perform the recovery of the Command Module and crew.

USS *Goldsborough* (DDG-20)

The guided missile destroyer USS *Goldsborough* (DDG-20) was commissioned at Puget Sound Naval Shipyard in Bremerton, Washington in November 1963. Guided missile destroyers are fast combatant warships designed to handle multiple missions. These include operating in support of carrier battle groups, surface action groups, amphibious groups and replenishment groups as well as providing naval gunfire support. In 1965, she was a member of the recovery force for *Gemini 4* and *Gemini 5*.

By the summer of 1969, she had completed four tours of duty off the coast of Vietnam in only five years. Her missions included providing gunfire support for ground operations and escorting attack carriers on Yankee Station in the South China Sea. In 1967 she participated in "Operation Sea Dragon," designed to interdict the North Vietnamese lines of supply into the Republic of Vietnam, and provided gunfire support along the DMZ. During this deployment, *Goldsborough* fired nearly 10,000 rounds of ammunition and avoided over 800 rounds of hostile fire without damage to the ship.

When not involved with the Vietnam conflict, she had also participated in Project UPTIDE with *Hornet*, researching fleet ASW techniques against Soviet nuclear powered submarines.

For the *Apollo 11* mission, her Commanding Officer was Commander Paul Lautermilch, a 1951 graduate of the U.S. Naval Academy. The Secondary Recovery Ship had two assignments. On July 16, while the spacecraft performed the TLI burn, her mission abort station (latitude $10°$ north, longitude $175°$ west) was located half way between Johnston Island and *Hornet's* TLI abort location near the Phoenix Islands.

Fig 6C—The USS *Goldsborough* is shown underway inside Pearl Harbor in August 1968. She was awarded the Naval Unit Commendation for exceptionally meritorious service in Vietnamese waters.

Fig 6D— Named after a river in Arizona, the USS *Hassayampa* was the third of six large Neosho-class fleet oilers built when the Cold War was heating up. The large quantity of supplies it carried enabled the Navy carrier battle groups to operate at sea for an extended period of time.

During the End-Of-Mission recovery phase on July 24, TF-130's plan called for *Goldsborough* to be stationed at the farthest end of the spacecraft re-entry corridor, northeast of Hawaii.

USS *Hassayampa* (AO-145)

Fleet oilers replenish petroleum products and ordnance to the fleet at sea using a process called "underway replenishment" (UNREP). The oilers transport bulk petroleum and lubricants from depots to the ships of a battle group. They also transport limited freight, mail and personnel to combatant ships and support units underway, enabling those ships to remain "on station" for as long as necessary.

The USS *Hassayampa* was commissioned in April 1955 and was home ported at Pearl Harbor for most of her career. She made numerous deployments to the Western Pacific, providing logistics for many Seventh Fleet operations, including the protection of Quemoy-Matsu from possible communist Chinese invasion in 1958. In 1965, *Hassayampa* supported Operation Market Time, a joint effort between the U.S. Navy and the South Vietnamese Navy, in an effort to stop the flow of supplies from North Vietnam into the south by sea. In 1966, *Hassayampa* served as a recovery logistics ship for the *Gemini 8* and *Gemini 9* flights. Shortly thereafter, she returned to the Far East and over the next five months, refueled 367 ships in support of Vietnam operations.

According to the TF-130 plan, under the command of Captain Jack E. Waits, she was to perform an underway replenishment for *Hornet* as the latter proceeded to the *Apollo 11* primary landing area. This ensured *Hornet* would have plenty of supplies to handle any contingency requirements that might arise.

USS *Arlington* (AGMR-2)

A 14,500-ton Communications Major Relay ship, USS *Arlington* (AGMR-2) was originally commissioned in July 1946 as the light aircraft carrier USS *Saipan* (CVL-48). She was mainly involved in pilot carrier landing qualification training, and aircraft transport duties. As a result, she helped transition the fleet to jet aircraft. In August 1966, after a lengthy and complex conversion, the ship was renamed USS *Arlington*, one of only two Navy vessels to ever be designated an AGMR. A floating relay station, *Arlington* carried ten 10,000-watt transmitters, fourteen 500-watt transmitters, sixty-five multi-channel receivers, and 1,200 pieces of teletype equipment plus special satellite terminals. She was equipped to provide naval commanders with the extensive command and control communications required to conduct a wide range of naval operations.

Arlington was stationed in the Far East area from August 1967 until late 1968, providing reliable message-handling facilities for ships of the Seventh Fleet in support of combat operations. In addition, she assisted ships in repairing and optimizing their electronic equipment. She did several tours of duty in the Tonkin Gulf area and was awarded seven campaign stars for this service.

Fig 6E—With her array of radio antennas and a satellite terminal, USS *Arlington* augmented the sparse DoD communications facilities in the southwest Pacific during the Vietnam conflict. She proved to be a great asset to the recovery forces for the first three Apollo lunar flights.

In December 1968, *Arlington* joined TF-130 as the primary communications relay point for the *Apollo 8* recovery operation, handling news media traffic between the USS *Yorktown* (CVS-10) and Honolulu. Only two weeks later, she was again operating off the Vietnamese coast.

In May 1969, she re-joined TF-130 to assist USS *Princeton* (LPH-5) with communications relay activities for the *Apollo 10* recovery operation. Immediately following this assignment, *Arlington* proceeded to Midway, where she provided communications support for the Nixon-Thieu summit conference on June 8. She then returned to the Vietnamese coast.

On July 7, she was ordered east for her third Apollo mission. She was added to the *Apollo 11* recovery force when the Navy learned that President Nixon and his staff planned to participate in the recovery activities. For one night, *Arlington* was scheduled to become a floating White House.

Before leaving Guam to join *Hornet* near Johnston Island, White House Communications Agency (WHCA) and Secret Service personnel embarked to create a secure berthing and communications environment for the President. Arriving in the recovery area on July 21, she tested her communications equipment and moved to Johnston Island the next day.

In addition to providing secure communications for the recovery forces and the President's staff, *Arlington* augmented *Hornet's* capabilities to ensure adequate support for the media, whether wire service or voice calls. During the *Apollo 11* recovery operation, her Commanding Officer was Captain Hugh Murphree.

Fig 7A—A C-1A Trader transport aircraft is launched from *Hornet*.

Chapter Seven
Primary Recovery Force: Units

Shipboard Units

The major units embarked on *Hornet* as part of the *Apollo 11* recovery force reflect a fairly typical configuration for lunar missions. The White House Communications Agency and Secret Service staffs are omitted from this description, but certainly had an impact on the operational environment. Based on previous Apollo missions, four major units were required to perform an effective operation.

Helicopter Anti-Submarine Squadron Four (HS-4)

During an Apollo recovery operation, the helicopter squadron was assigned multiple missions. Of course, they had to execute their normal ship-at-sea operational responsibilities. For instance, they served a "plane guard" role in case of a launch catapult failure or other accident caused an aircraft to ditch into the sea. Their space mission activities included locating the Command Module splashdown point, deploying UDT personnel to assist with astronaut and CM recovery, retrieving the astronauts and ferrying them back to the Primary Recovery Ship, and providing a photography platform to capture still and movie images of the recovery process. Helicopters not directly involved in the astronaut or Command Module retrieval operations carried grappling hooks to snag the CM parachutes floating in the water while also observing the general area for sharks or other hazards to the UDT swimmers.

Fig 7B (above)—The Sea King provided a very stable platform from which to conduct either Anti-Submarine Warfare operations or Search and Rescue activities. For ASW operations, the crew dipped a sonar ball into the water and listened for undersea craft in the area. For SAR, they used the Billy Pugh net (shown) to hoist a person into the aircraft from the sea or a disabled vessel.

HS-4 was commissioned in June 1952 at U.S. Naval Auxiliary Landing Field, Imperial Beach, California. Over the years, the squadron flew several different types of helicopters, including the HUP-2 Retriever, HO-3S Dragonfly, H-19 Chickasaw, SH-34 Seabat and SH-3 Sea King. HS-4 was the first Anti-Submarine Warfare (ASW) helicopter squadron to deploy aboard an aircraft carrier. In 1961, it became the first squadron in the Naval Air Forces Pacific Command to achieve around-the-clock ASW capability, earning HS-4 the title "Black Knights."

The squadron was initiated into the "Tonkin Gulf Yacht Club" during a 1966 WESTPAC cruise off South Vietnam. During this deployment, HS-4 pilots and aircrewmen rescued twenty-four downed airmen, the largest number recorded by any ASW squadron during the Vietnam conflict. By 1968, when they performed the *Apollo 8* recovery, HS-4 was equipped with the Sea King helicopter.

The Sikorsky SH-3 Sea King became operational with a few Navy squadrons in 1961. It was a state-of-the-art, all-weather, night and day, Anti-Submarine Warfare he-

Fig 7C (right)—The Sea King is a very large helicopter compared to previous rotary winged aircraft. This frontal view of Helicopter #66, taken on the flight deck of *Hornet*, clearly shows the boat-shaped hull and landing-gear sponsons that allow it to float on water in an emergency.

licopter. Each of its two General Electric T-58 gas turbine engines produced 1,400 shaft horsepower. The maximum endurance was over six hours with a maximum range of about 700 nautical miles. At one point, the SH-3D held the world speed record for helicopters at 160 mph. Its rescue hoist could handle up to a 600-pound capacity.

The adaptable platform was designed for ship-based operations—its five main rotor blades as well as its tail rotor section could be folded for easy stowage. The Sea King's primary missions were Anti-Submarine Warfare (ASW) and Search and Rescue (SAR), but it was also used in general personnel and cargo transport, communications, executive transport and airborne early warning roles. It served as the primary fleet helicopter of the Navy for over thirty years.

In preparation for the *Apollo 11* mission deployment, HS-4 practiced with UDT personnel in San Diego Bay. One half of the full squadron complement of sixteen helicopters was assigned to the recovery mission with Commander Donald Jones. These eight Sea Kings were configured in search-and-rescue mode, eliminating their usual ASW equipment. During SAR missions, the flight crew consisted of two pilots and two aircrewmen.

HS-4 was the recovery squadron for the first five lunar orbit and/or landing flights, the all-time record, for which the squadron was awarded a Meritorious Unit Commendation. Commander Don Jones was the Commanding Officer during the *Apollo 8*, *10* and *11* recoveries while Commander Charles Smiley was in charge during *Apollo 12* and *13*.

During the *Apollo 8* recovery in December 1968, an SH-3D Sea King helicopter (Navy bureau number 152711) retrieved the astronauts in full view of the TV cameras. The two-digit squadron number painted on the fuselage was 66, and it became a widely recognized public symbol for "lunar mission accomplished." Each time it successfully completed a recovery mission, a spacecraft

Fig 7D—A dramatic photo of Helicopter #66, Navy Bureau Number 152711, in flight near Oahu, Hawaii in 1969.

decal was added onto the fuselage under the cockpit windows. The Navy switched over to a three-digit squadron numbering scheme a year later and this helicopter became number 740. However, to maintain public recognition, HS-4 repainted the number 66 on its fuselage just before making each successive Apollo recovery. BuNo 152711 was manufactured at the Sikorsky factory in Stratford, Connecticut in April 1967 and became the Black Knight's third Sea King when it was delivered later that month. The primary recovery helicopter for *Apollo 8*, *10*, *11*, *12* and *13*, Helicopter #66, crashed off the coast of southern California during a training mission in June of 1975 and remains on the sea floor.

Airborne Early Warning Squadron VAW-111

Fig 7E—The E1-B Tracer was a variant of the venerable S-2 Tracker and was designed as an airborne early warning (AEW) aircraft.

When embarked on an aircraft carrier, the normal VAW squadron mission was to provide airborne early warning of potential attacks (i.e., act as long range eyes for the fleet) and provide intercept command and control in the event of an air or submarine attack. For an Apollo recovery assignment, the primary mission of these airborne radar platforms encompassed three activities. They aided in identifying the location of the Apollo Command Module during its re-entry, acted as a communications relay point of the astronaut's voice frequency to the Primary Recovery Ship and the TF-130 headquarters in Hawaii, and assisted with air space control of the recovery helicopters during night or instrument flight operations. Having extended radar coverage of the Command Module was particularly important because of the normal communications blackout that occurred during this stage of its flight. For the *Apollo 11* recovery, VAW-111 was given the additional responsibility of tracking the flight of the Presidential helicopters to and from the recovery area.

VAW-111 flew the E-1B "Tracer," a twin engine, folding wing, carrier-based plane built by Grumman Aircraft in the late 1950s. It was based on the S-2 Tracker anti-submarine warfare aircraft, but was modified to have twin tails. It was originally called the WF-2, and soon garnered the nickname of Willy Fudd. The immediately recognizable feature of the E-1B was the massive radome above the fuselage containing a state-of-the-art Hazeltine AN/APS-82 radar. This radar introduced many technological advances, including a stabilized antenna and Airborne Moving Target Indicator (AMTI), allowing the radar operator to detect low-flying targets against the clutter of radar reflections from the surface of the ocean. The Tracer normally carried a crew of four, including a pilot, copilot, radar controller and a flight technician.

VAW-111 Detachment 12 was assigned to provide four E-1Bs with trained crews and support personnel to the *Apollo 11* recovery mission. The Tracers extended the radar coverage beyond the horizon, allowing *Hornet* to manage a wider airspace than normal.

Fig 7F—A VR-30 C-1A Trader, configured for its USS *Hornet* Carrier-Onboard-Delivery (COD) assignment, shown in flight.

Fleet Logistics Support Squadron VR-30

The primary mission of a fleet logistics support squadron was to carry various pieces of cargo and personnel between shore facilities and carrier battle groups at sea. They moved men and material among shore-base naval stations, training centers and supply bases. For the *Apollo 11* recovery mission, the key responsibilities were to ferry VIPs to and from *Hornet* for the major events and to transport the Moon rocks from *Hornet* to a major shore-base airfield as soon as they could be packaged for flight.

In October 1966, Fleet Tactical Support Squadron 30 (VR-30) was established at Alameda Naval Air Station, equipped with Convair C-131 Samaritans and Grumman C-1A Trader aircraft.

In November 1966, VR-30 made their first C-1A Carrier Onboard Delivery (COD) arrested landing on an aircraft carrier at sea. To support the COD mission, the C-1A transport aircraft had a flight crew of two and the passenger cabin configured to hold nine seats plus a storage hold. It carried cargo, such as mail, and up to eight passengers, plus an aircrewman in the seating area. The squadron was awarded the Meritorious Unit Commendation for exemplary service from January through November 1967. In 1969, squadron C-1A's and crews operated from Danang, Republic of Vietnam, in support of the Navy's Task Force 77 operating in the South China Sea.

For the *Apollo 11* mission, the C-1A cabins were placed in their "Charlie" configuration, which provided four passenger seats plus a cargo cage. From 1968 to 1973, VR-30 COD detachments operated aboard various carriers in support of recovery operations for *Apollo 10*, *11*, *12* and *16*.

Underwater Demolition Teams 11 and 13

During the Vietnam War, three 108-man teams were headquartered at the Naval Amphibious Base Coronado and rotated through on a scheduled basis. At any point in time, one team was in the combat theater, one team was recovering from a combat tour and the third was working up to deploy to the western Pacific. Because lunar flights landed on the Mid-Pacific Line, the brunt of the Apollo recovery operations fell on these teams.

In December 1968, after *Apollo 8* orbited the Moon, frogmen from UDT-12 ensured a safe and speedy recovery. However, by early 1969, UDT-12 was deployed in South Vietnam and conducting combat operations. UDT-13 was at Coronado but undergoing intensive training for its upcoming WESTPAC rotation. UDT-11 had just returned from Vietnam, and became the "duty team" at Coronado, which coincided with the Navy's selection of personnel to perform both the *Apollo 10* and *Apollo 11* recoveries. In May 1969, *Apollo 10* completed its mission as a dress rehearsal for a lunar landing, and the UDT-11 swimmers effected a flawless recovery of crew and craft.

Typical spacecraft recovery missions were assigned three officers and eight enlisted men, divided into three teams. The plan called for two teams to be in flight during the Command Module re-entry, while the third acted as backup waiting in a helicopter on the deck of the Primary Recovery Ship.

All personnel assigned to spacecraft recovery duty underwent many hours of specialized training under NASA guidance. Even after splashdown, the CM contained several hazards, including explosive pyrotechnic devices, toxic fumes and possible sudden chemical fires due to the fuel used by the spacecraft reaction control thrusters. In addition to the standard UDT equipment carried by each swimmer, the three-man recovery team was equipped with one spacecraft flotation collar, three eight-foot diameter sea anchors and a seven-man life raft.

Fig 7G —Prior to embarking on the *Apollo 11* mission, the UDT team practiced recovery techniques in San Diego Bay for four weeks. A UDT swimmer has just leaped from the hovering helicopter while his teammate waits in one of the rafts connected to the boilerplate CM.

The principal objectives of the UDT swimmers during a spacecraft recovery normally were:

- Clear the huge parachutes away from the Command Module area
- Secure the floating spacecraft by attaching a sea anchor, flotation collar and two rafts
- Assist the astronauts in exiting the CM into a raft for pickup by a helicopter
- Prepare the spacecraft for retrieval and remain with it until it was hooked to the ship's B&A crane.

These basic techniques had been well-honed during the Gemini and early Apollo missions.

Apollo 11 carried the significant added complexity of maintaining a quarantine against possible lunar pathogens. The "Moon germ" issue required an additional officer to handle the decontamination requirements. The latter required the use of special quarantine procedures and equipment such as the Biological Isolation Garment suits. The Command Module needed to be decontaminated, the astronauts needed to don their BIG suits and all four men who had been in contact with the CM (three astronauts and one UDT swimmer) needed to be scrubbed with a decontaminate as well.

The UDT functions were expanded to include:

Fig 7H—Clancy Hatleberg, clad in his BIG suit (rather than a wet suit), is shown handing a piece of equipment to one of the *Apollo 11* astronauts during training in the Gulf of Mexico. He was the only UDT person who trained with the astronauts prior to the actual mission.

- Initial quarantine and decontamination of the astronauts before helicopter recovery
- Decontamination of the Command Module before retrieval by the PRS.

Lieutenant Clancy Hatleberg was designated the biological decontamination ("decon") swimmer for the *Apollo 11* mission, a completely new role in the spacecraft recovery program. It was his responsibility to perform the new decontamination procedures prior to the astronaut's departure from the splashdown scene.

The primary and standby teams for *Apollo 11* were from UDT-11, while the secondary team was from UDT-13. The selection of the primary team was based on its successful involvement in the recovery of *Apollo 10*.

Non-Shipboard Support Units

Other DoD units not stationed onboard *Hornet* provided key support as well. The Aerospace Rescue and Recovery Service (ARRS) teams were placed under the control of TF-130 while the Apollo Range Instrumentation Aircraft (ARIA) were directed by the Air Force Eastern Test Range (AFETR) located at Patrick Air Force Base in Florida.

USAF Aerospace Rescue and Recovery Service

Pararescue jumpers (aka PJs) of ARRS were highly trained rescue professionals. They were precision parachutists, with medical technical expertise, trained in survival techniques and SCUBA certified. PJs operated in three-man teams and exited their aircraft at about 1,000 feet. Each PJ wore a wetsuit and carried more than 170 pounds of equipment. Personnel assigned to spacecraft recovery duties underwent many hours of training similar to that of the UDT swimmers. For an Apollo recovery mission, each man also carried an accessory kit containing a radio, snorkel, flashlight and Apollo inter-phone device.

The PJs would attach the flotation collar, a sea anchor (using a reserve parachute), and a six man raft. These activities ensured the safety and comfort of the flight crew until a Navy ship arrived. This airborne rescue capability proved its value to NASA space programs during the Mercury *Aurora 7* flight in 1962 and again during the *Gemini 8* mission in 1966.

Some HC-130H aircraft were outfitted with the Fulton surface-to-air-recovery system, also know as the sky-hook. The recovery kit was designed to rescue one or two men. A kit was dropped to the person to be recovered. He then put on a special harness and inflated the balloon, which then raised the attached lift line into the air. An HC-130H snagged the line with a V-shaped yoke protruding from its nose, and the individual was reeled on board. Red flags on the lift line guided the pilot during daylight recoveries; lights on the lift line were used for night recoveries. In the end, the system proved impractical for most rescue purposes, but was very useful for special operations.

Fig 7I (opposite)—The Lockheed C-130 Hercules shown in its ARRS "rescue" configuration with the Fulton surface-to-air-recovery sky-hook device mounted on the nose and Apollo direction-finding pod on top of the fuselage.

Fig 7J (below)—This navigation chart was used by Major Charles Hinton of the ALOTS aircraft to plot and record information about the CM's re-entry. The notations in red—start, max, end—refer to the fireball created as the CM burned deeper into the atmosphere, starting at 400,000 feet.

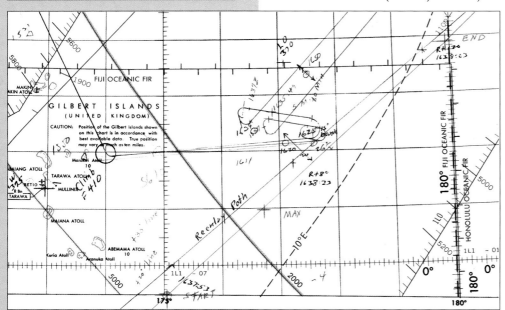

The commander of the Pacific ARRS Center in Hawaii was Colonel Thomas Shockley. Prior to the launch, he assigned four HC-130H ARRS aircraft to TF-130 for the *Apollo 11* mission. Two Hercules from the 79th ARRS were staged out of Andersen Air Force Base in Guam and supported the Trans-Lunar Injection burn event on July 16. Two others from the 76th ARRS were staged out of Hickam Air Force Base in Hawaii and handled the End-Of-Mission recovery on July 24. The latter two had call signs *Hawaii Rescue 1* and *Hawaii Rescue 2*. They were positioned roughly 190 miles from *Hornet* along the Command Module re-entry path; one was posted up-range and while the other was down-range. Both orbited their assigned locations at 10,000 feet, ready to track the CM after it came out of the S-band blackout period, with their PJs ready to jump if needed.

USAF ALOTS & ARIA

In support of the *Apollo 11* mission, AFETR assigned three Apollo Range Instrumentation Aircraft (ARIA) aircraft, although they did not travel together. The ARIA needed to be "on station" by July 16 to cover the TLI burn, 1,600 miles southwest of Hawaii. The plan called for them to fly from bases in Australia, Guam and Hawaii, since it was not known in advance on which orbit of the Earth the TLI burn would occur. They remained in the Pacific area until eight days later to provide communications relay functions during the atmospheric re-entry of the spacecraft.

An ALOTS aircraft was assigned to film the atmospheric re-entry of the Apollo spacecraft on July 24. Flown by Colonel Oakley Baron, the ARIA Squadron Commander, the ALOTS aircraft left Patrick AFB on July 19th and was pre-positioned on Wake Island by July 22. Of all the re-entry aircraft, this one had the smallest window of opportunity to perform its task and was provided with sophisticated navigation information before the mission. As this chart shows, the aircraft had to be at a precise location, exact altitude, and specific course at almost the exact second to catch and film the speeding *Columbia* as it tore through the upper atmosphere.

Fig 7K—USAF Eastern Test Range ARIA number 375, with its unique nose-mounted radome, is shown in flight during the Apollo era

Technical Note:
Navigational Aids for Locating the Command Module

Because a recovery operation might occur at night or in other limited-visibility situations, NASA placed two VHF transmitting antennas and a flashing light on top of the Apollo Command Module. These locating aids were activated by the spacecraft's crew just after the main parachutes deployed around 10,000 feet.

USAF ARRS HC-130H aircraft were outfitted with AN/ARD-17 radio direction finding-equipment; the antenna for which was housed in a radome on top of the fuselage, forward of the wings. This allowed them to track the Command Module S-band signal during the re-entry phase of the flight, and obtain reception of the VHF recovery beacon after the CM had landed.

Navy recovery helicopters were outfitted with a basic radio direction-finding system called Search and Rescue and Homing (SARAH). A SARAH yagi-type antenna was mounted onto the landing gear strut

Fig 7L—HS-4 Executive Officer CDR Chuck Smiley gently settles Sea King #66 onto *Princeton*'s flight deck with the *Apollo 10* astronauts inside. The key Apollo recovery components are clearly visible as noted along its starboard side.

on both sides of the Sea King and a receiver was placed in the normal search and rescue equipment bay. As the helicopter flew in the general direction of the spacecraft's transmitter beacon, the relative strengths of the two signals would indicate how to alter the flight course to arrive at the correct location—left, right or straight ahead. When the signal strength dropped dramatically, the helicopter had flown over the CM and was heading away from it.

This close-up photo, taken during the *Apollo 10* recovery, also shows the black-painted rescue hoist assembly that keeps the cable from rubbing against the side of the helicopter fuselage. During the actual astronaut retrieval event (not SIMEXs), a 70mm film camera was also mounted just behind the cargo hatch on the recovery helicopter to provide an overhead record of the hoisting operations.

Just below the cargo hatch was a special uprighting sling assembly. One end of it was attached to a bomb shackle (hard point) on the helicopter, then a portion of it was taped along the side of the fuselage and ran into the cargo hold where the rest of the line was coiled. Although the Earth Landing System (ELS) of an Apollo spacecraft was rigged for the CM to land upright (i.e., apex up), the action of wind or waves could overturn it. If, after landing, a Command Module turned apex-down (Stable 2), and its inflatable uprighting bags malfunctioned, the winch operator could lower this seventy-five-foot, half-inch nylon line and hook assembly to a UDT swimmer. He, in turn, would latch it onto the CM and the helicopter could gently lift it into an upright position.

Fig 8A—HS-4 helicopter aircrew prepares to drop rafts to the UDT team that has already attached a flotation collar to the boilerplate.

Chapter Eight
Pre-Flight Preparations

On May 13, the USS *Hornet* returned to Long Beach from its third and final Vietnam deployment, which had lasted seven months. Due to a normal Navy personnel rotation cycle, Captain Carl Seiberlich became its new commanding officer ten days later. Commander Chris Lamb was advanced from his previous position of ship's Navigator to being the Executive Officer. On June 5, *Hornet* was formally designated the Primary Recovery Ship (PRS) for *Apollo 11*. Captain Seiberlich automatically became the Commander of all Recovery Forces in the primary landing area. In preparation for this mission, *Hornet* was given a three-week overhaul.

While the basic spacecraft recovery procedures had matured, *Apollo 11* was the most complex recovery operation in the history of the American manned space program for three main reasons—pathogens, President and press.

Because it was the first lunar landing flight, the scientific community was worried about lunar pathogens possibly contaminating the Earth. Therefore, the astronauts, spacecraft, Moon rocks and all the equipment that were inside the spacecraft would be subject to the new quarantine procedures. Unlike all previous NASA missions, there would be no astronaut reception event on the flight deck.

For the first and only time, the President of the United States would be an eyewitness to the recovery operation. The presence of President Nixon, and other high-ranking officials, greatly increased the complexity of handling media, communications and security arrangements. His needs were "overlaid" on top of an already complex mission.

Finally, worldwide media interest was intense. Reporters from print, radio and TV news agencies were vying for credentials to be on *Hornet* for this mission.

Extensive planning meetings were quickly held with the many organizations involved, such as the White House Communications Agency (WHCA), NASA, Navy, Air Force, media, etc., to ensure effective communication and cooperation. Senior staff of the USS

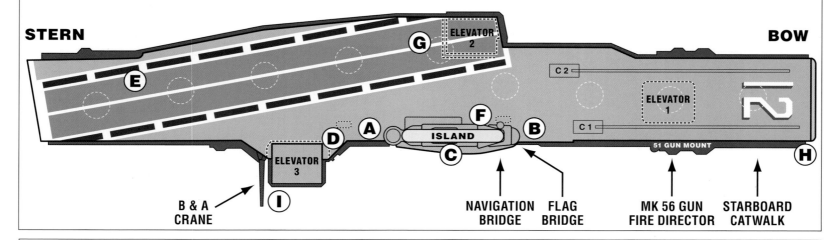

Fig 8B—This graphic shows the *Hornet* flight deck layout and several of the locations where equipment was located or key recovery activities took place.

Princeton provided valuable information about their experience with the recovery of *Apollo 10* and the follow-on evaluation of quarantine procedures.

It immediately became apparent that space allocation was a major challenge. During the planning sessions, it was decided that Hangar Bay #2 would be the central point for all recovery activities. The main consideration was ease of accessibility, both from the port side aircraft elevator down which the astronaut helicopter would be lowered after landing on the flight deck and the starboard side aircraft elevator on which the spacecraft would be placed after retrieval from the ocean. It also allowed Hangar Bay #1 to remain available for handling mission aircraft, taking them off the flight deck at night or in rough weather. Hangar Bay #3 remained usable for aircraft maintenance as required throughout the recovery operation and also provided storage space for the NASA boilerplate and UDT equipment.

While in Long Beach, a wide range of media equipment for TV broadcasting and radio communications was loaded onboard. On the flight deck, General Electric's first ever satellite communications

Fig 8C (above)—Technicians from General Electric are setting up their satellite antenna on the flight deck prior to covering it with the white "bubble" dome that protected it from salt spray. Once it was operational, the satellite system was used for broadcasting live TV and other traffic to the U.S. mainland.

Fig 8D (left)—*Hornet* crewmen guide one of the ABC-TV broadcasting vans onto elevator #3 from where it was towed into the hangar deck and tied down for the entire journey.

antenna for color TV transmissions (protected by a white bubble dome) was located behind the ship's island. The Mutual Broadcasting radio van was located midway along the starboard side of the island with multiple dipole antennas arrayed high on the island superstructure. To interconnect the aircraft carrier with other aircraft and ground stations in Hawaii, the Navy installed a UHF-based TACSAT satellite terminal and antenna hut just in front of the island. NASA placed a VHF-based ATS-1 satellite terminal behind the island to handle VIP voice calls into the MQF.

On the hangar deck, three large ABC-TV News broadcasting trailers were tied down at the front of Hangar Bay #2. The Voice of America loaded broadcasting equipment onboard to be certain this event was heard behind the Iron Curtain. Along with this specialized equipment came 120 technicians from NASA, GE, ABC and Western Union International (WUI). The ship's engineering department had to create a number of new electrical and telephone connections to support these systems. Unfortunately, one item that could not be installed in time was the TRANSIT terminal used during the *Apollo 10* recovery that would permit more precise ship navigation via the advanced Navy Navigation Satellite System.

On June 26, *Hornet* departed Long Beach for San Diego where she embarked the airgroup maintenance and support personnel. The next day, after *Hornet* left San Diego for Pearl Harbor, the aircraft were flown aboard following customary Navy procedure. This included eight SH-3D Sea King helicopters from HS-4, four E-1B Tracer communications relay aircraft from VAW-111, and two C-1A Trader COD aircraft from VR-30 to augment *Hornet's* single COD. Over 250 men comprised this specialized recovery group.

On July 2, *Hornet* moored at Pier Bravo at Naval Station Pearl Harbor and was immediately "chopped" (temporarily assigned) to TF-130. Shortly thereafter, some of the NASA recovery equipment, including a boilerplate capsule (BP-1218), was hoisted aboard.

Fig 8E—The initial NASA recovery equipment, including the boilerplate capsule and the ATS-1 satellite terminal, is brought to the pier and prepared for hoisting aboard by the ship's B&A crane.

Along with the UDT detachment, numerous members of the news media boarded the ship. The TV broadcast media included Ron Nessen of NBC, Dallas Townsend of CBS, and Keith McBee of ABC. Don Blair, a well-known announcer for the Mutual Broadcast System handled the live radio coverage. Commercial television broadcasts provided by ABC were transmitted via GE under contract to Western Union International. *Hornet* set up a general working area for members of the media in the island superstructure, on the same passageway as the Combat Information Center and Radio Room. The press center was located in the Flag Operations Office, a press briefing area in the adjacent War Room and a teletype area in the Debriefing Room. This greatly simplified coordination, since these spaces were also adjacent to pilot Ready Room One where the HS-4 and UDT command center was located.

Captain Seiberlich wasted no time in establishing crew responsibilities and ensuring adequate training. There were twenty-seven discrete recovery tasks that had to be performed from the moment of splashdown to the time *Hornet* could steam away from the End-Of-Mission landing area. In addition to the astronaut recovery teams of HS-4 and UDT-11, six shipboard units were created and assigned responsibilities for the retrieval process of the spacecraft:

- Command Module retrieval team (crane operators, etc)
- Security team
- Radar acquisition and tracking team
- Life boat team
- Records and reports team
- Crowd control team

Fig 8F—UDT swimmers played the role of astronauts, with one of them being hoisted in the Billy Pugh rescue net to simulate the astronaut retrieval process.

The slogan, *Hornet Plus Three*, was specifically chosen to depict the safe recovery of the three astronauts as well as the return of all recovery and media personnel from the operation.

During the period of July 7-9, *Hornet* conducted nine Simulated Recovery Exercises (SIMEX) in local Hawaiian waters. Different recovery scenarios were simulated to work out various contingency procedures. The SIMEXs involved a wide range of personnel and equipment; six were performed during daylight and three at night. Captain Seiberlich wanted at least three people trained for each major task, ensuring there was no single point of failure. Unit commanders determined how to creatively use various ship assets to assist in the Command Module recovery. For instance, by positioning the ship's motor whaleboat in the vicinity of the CM with personnel equipped with walkie-talkie radios and still cameras, they could improve safety, communications, and the quality of photographs. The ship's MK56 gunfire control radar for the number 51 anti-aircraft gun mount near the bow proved useful in obtaining accurate range and bearing information during the ship's approach to the Command Module by locking on to a helicopter hovering over it.

Other exercises verified the operation of the wide array of military and civilian communications systems that carried links to NASA's Mission Control center in Houston and the TF-130 Recovery Control Center in Kunia. These included UHF on-scene surface and HF air-to-ground voice radio plus the HF/SSB radio, Tactical Communications Satellite (TACSAT) and Applied Technological System (ATS-1) communications systems. Western Union verified the viability of the links connected via INTELSAT IV for the TV and radio media. Some issues were discovered that needed to be worked out, such as TV interference caused by *Hornet's* air search radar. A modern aircraft carrier, in the middle of the Pacific Ocean, is a hostile environment for normal land-based communications systems and many alternative strategies were tested.

On July 9, *Hornet* returned to Pearl Harbor to complete the final loading of supplies. The next day, two MQFs and other quarantine equipment, including fifty-five one-gallon containers of sodium hypochlorite, were loaded aboard. The rest of the NASA personnel, press and

Fig 8G—A Navy tugboat eases a barge carrying special quarantine equipment up a channel in Pearl Harbor. Clearly visible onboard are the two MQFs with NASA logos and piles of other equipment needed to ensure a successful recovery.

TV representatives walked up the main brow from the pier and were escorted to their staterooms. President Nixon's naval aide, Secret Service agents and WHCA personnel met with *Hornet* staff to establish logistics, security and communications arrangements for the President's visit.

On July 12, *Hornet* departed for its Mid-Pacific Line launch abort station, just below the equator and 1,600 miles southwest of Hawaii. *Goldsborough* headed for its assigned secondary launch abort site. That afternoon, Captain Seiberlich held the first onboard press conference, which continued at the rate of two a day during the entire mission. Guest speakers included the commanding officers of *Hornet*, HS-4, and UDT-11 as well as other key Navy and NASA personnel. Press copy was transmitted to the continental U.S. by WUI twice a day, using seven teletype channels. The Navy was not involved in the handling of media reports.

The next day, two US-2C utility transport aircraft from squadron VC-1 stationed at Barbers Point Naval Air Station near Honolulu made their inaugural run to *Hornet* in preparation for ferrying dignitaries and other media personnel between Hawaii and the ship.

Hornet provided entertainment for the ship's crew and visitors through various means. Two feature movies were shown each night in the officer's wardroom, and a ship's

Admiral Seiberlich recalls: [We had] a couple hundred newspaper, TV and radio personnel reported onboard to cover the event. I briefed them about conserving water, since the ship's boilers had first preference on the desalinated water supply to keep the ship moving. But, within forty-eight hours of their arrival, we had run out of water. We explained to them how to take a "Navy shower," and after that, we didn't have a problem.

In an effort to get the best photo, some of the media got into fights with each other so we made chalk outlines on the deck to show each one where to stand during the recovery operation.

Fig 8H—SIMEX activity also involved recovering the boilerplate, requiring the giant aircraft carrier to slowly creep alongside the tiny spacecraft so a line could be tossed to the UDT team without endangering the swimmers.

newspaper was provided every day. The COMNAVAIRPAC band played two concerts a day, one after lunch and another after dinner. A talent show was held two days prior to splashdown. Bridge, pinochle and acey-ducey tournaments were arranged for both civilian and Navy personnel.

Each of the various primary and backup recovery teams showed constant improvement as nine more SIMEXs were made prior to the *Apollo 11* launch. Unusual anomaly situations were injected into the process. Night and early morning operations were not uncommon and exercises were held in different sea states to train personnel and verify procedures regardless of surface weather conditions. Relatively minor changes in weather conditions could disrupt a recovery, such as a shift in wind direction that could blow the Command Module into the path of the aircraft carrier. The ship's quartermasters worked out various navigational solutions for both portside and starboard side approaches when recovering the CM. Since the ship's massive Boat & Aircraft (B & A) crane was located on the starboard side, a port side approach required the use of a mobile crash crane, called a Tilley, to effect the Command Module retrieval; that procedure was rehearsed as well. The medical department trained for more than the usual contingencies of normal sickness and injury of crewmen or civilians. It also had to prepare for the possible contamination of personnel due to a break in the quarantine process and the spread of lunar pathogens onboard.

Members of the press, especially the TV crews, used the SIMEX routines to determine the best vantage points from which to report on the recovery activities. Several TV cameras were hard-mounted in key locations high on the island superstructure as well as on temporary platforms on the flight deck and in Hangar Bay #2. The TV media desired one of the heavy cameras to be mobile so it was attached to an aircraft tug to be pulled about the flight deck to gain

Fig 8I—The "mobile" TV camera on the flight deck, attached to an aircraft tug, zooms in on one of the permanently mounted TV cameras located directly over the Navigation Bridge on the island superstructure. The Navy band is preparing for a rehearsal on the flight deck.

the optimum vantage point to record different activities. Mutual Radio announcer Don Blair staked his claim to a place on the signal bridge that had great visibility.

Early on July 15, *Hornet* crossed the equator, much to the delight of King Neptune. Since ancient times, mariners crossing the equator for the first time have participated in a colorful ceremony inducting them into the kingdom of Neptunus Rex. Before the crossing, they are referred to as "pollywogs," and are subject to various types of hazing and humiliation. Members of the media and NASA personnel were swept up into the festivities along with the ship's crew. However, Captain Seiberlich decided to end the program before its usual conclusion. So many pollywogs were getting greasy hairdos and rotten food baths that getting them clean would have taken too much fresh water. By the end of the day, over 600 people had become "Trusty Shellbacks."

Hornet continued steaming south at fifteen knots. In the late evening of July 15, she arrived at her assigned launch abort station (latitude 3° south, longitude 165° west), about two hundred miles east of the Phoenix Islands. All hands eagerly awaited word of the launch of *Apollo 11*, scheduled about five hours from then.

Fig 8J—King Neptune presents himself and his court to Captain Seiberlich on the flight deck as *Hornet* crosses the equator. Several hours of colorful mayhem ensued.

Fig 9A—President Nixon descends the steps of *Marine One* aboard USS *Arlington*.

Chapter Nine
Preparing for the President

President Nixon was clearly swept up in the drama of the first lunar landing mission, not just for the sake of history but the prestige it bestowed on America. He had been in the White House only a few months when *Apollo 10* orbited the Moon. He was grappling with ways to wind down the war in Vietnam and to redefine America's role in Asia. This incredible scientific achievement provided the Nixon administration an opportunity to make progress on some tricky geopolitical issues.

Astronauts from previous flights had received a congratulatory telephone call from the American president while onboard the Primary Recovery Ship. None had ever been personally welcomed back. President Nixon decided in early 1969 to be present at the splashdown event, however, it was kept secret until just prior to the mission. Only on July 21, did President Nixon publicly announce he would add the Apollo recovery to the front end of his upcoming worldwide trip. Following the recovery, he planned to visit with the leaders of many Asian countries, including the Philippines, Indonesia, Vietnam, Thailand, India and Pakistan, to discuss the future U.S. role in Vietnam and Asia. His return to Washington, D.C. was via Europe, where he and Romania's leader, Nicolae Ceausescu, would discuss how to open a dialogue with Communist China.

When *Hornet* was selected as PRS, Captain Seiberlich was led to believe Vice President Spiro Agnew would be the senior civilian executive for the recovery. A NASA official quietly passed him the information just before *Hornet* left San Diego for Hawaii that the President would attend instead. Needless to say, the pressure for a picture-perfect recovery was greatly increased. Not only was Nixon the leader of the free world and Commander-in-Chief of all U.S. military forces, he was a former naval officer. Additional complexity was added by the two dozen members of the White House Press Corps covering the President's trip who were scheduled to arrive only one day before splashdown.

The switch also presented Captain Seiberlich with his first crisis. Previously, he had made a commemorative baseball hat for Vice President Agnew. Now, he needed one with the word "President" on it, but no one knew Nixon's hat size. He had the ship's supply officer order a number of caps made in different sizes, to be delivered upon the *Hornet's* arrival in Pearl Harbor.

The naval aide to President Nixon, Lieutenant Commander Charles Larson, was the "trip officer" for this part of the President's worldwide tour. He had to coordinate with many organizations including the Secret Service, White House Communications Agency, NASA, the Navy and the Air Force to ensure that all necessary preparations were made for the presidential visit. Two weeks before the launch of *Apollo 11*, he traveled to Hawaii, Guam and Johnston Island to meet with senior military officials.

The intended *Apollo 11* splashdown target point was several hundred miles southwest of Johnston Island. This tiny Pacific atoll, roughly 825 miles west-southwest of Honolulu, was the location of America's final atmospheric test of a nuclear weapon in November 1962. During the Apollo recovery in 1969, it held a secret U.S. military base then involved with Thor anti-satellite missile tests. It was the closest land mass to the primary spacecraft landing area with an airfield able to handle large jets, and became the staging area for logistical support of the President's visit.

For long flights in 1969, President Nixon flew on a USAF VC-137C that was designated *Air Force One* while he was onboard. This aircraft was a specially modified Boeing 707 with the tail number 26000. It entered service during President Kennedy's administration. For short flights, the President flew on a USMC Sikorsky VH-3A, whose normal call-sign was *Nighthawk*, but was re-designated *Marine One* when he was onboard. These helicopters are part of the U.S. Marine Corps HMX-1 squadron stationed at Quantico, Virginia.

Fig 9B—One of the HMX-1 helicopters is shown being off-loaded from a USAF C-133 cargo plane at Johnston Island after the flight from Andrews AFB.

The normal method of bringing visitors onboard an aircraft carrier at sea is by using the C-1A Trader COD aircraft, and making a cable-arrested landing on the flight deck. The Secret Service felt the landing of this aircraft, and the ensuing catapult launch required for the President's departure, constituted an unacceptable risk to the chief executive. Another factor involved the extensive security background checks performed on all HMX-1 personnel. There simply was not enough time to fully vette the Navy's executive transport pilots. The decision was made to use the normal HMX-1 helicopters to ferry Nixon and his party to *Hornet*.

During the initial planning stages, it was clear the Navy TF-130 recovery forces would be fully engaged in *Apollo 11* tracking and recovery activities. Arrangements were made to have additional Navy ships positioned between Hawaii, Johnston Island and *Hornet*, creating a "lily pad" effect in support of the HMX-1 flights to and from the splashdown area. The USS *Arlington* was designated the "at-sea base" for the presidential party. The plan called for *Arlington* to provide overnight berthing and aircraft refueling before the HMX-1 helicopters flew over to *Hornet* on splashdown day. Nixon was familiar with the *Arlington* officers and crew, since they had provided communications support for the recent summit conference on Midway Island with President Thieu of Vietnam.

On July 16, as the Saturn V rocket was launching *Apollo 11* from Kennedy Space Center in Florida, two USAF C-133s departed Andrews Air Force Base in Maryland, each carrying one HMX-1 helicopter. Both arrived at Johnston Island on July 18, where their crews immediately began reassembling them to flight status. A third Marine CH-53 helicopter was "borrowed" from an air unit on Hawaii and used for logistical support. On July 22, the Marine pilots flew to *Arlington* for carrier landing practice, since this was not something they normally did for presidential flights. They also carried support personnel and cargo in advance of the President's visit.

Fig 9C—President Nixon has just descended the steps of *Air Force One*. After being greeted by Navy Admiral John McCain and base commander USAF Lt Col C.E. McPhee, the party walks toward the HMX-1 helicopters. Behind Nixon is his naval aide LCDR Charles Larson.

President Nixon started his worldwide tour with a stop in San Francisco early on July 23. Nixon told a street crowd of several thousand people the Moon landing had generated a burst of good-will around the globe.

After flying from San Francisco to Hawaii, he and a small group of dignitaries then detoured on to Johnston Island. Nixon had visited it while returning home from wartime service in the South Pacific when he was a Lieutenant Commander in the Navy. Members of the party included Secretary of State William Rogers, National Security Adviser Henry Kissinger, NASA Chief Administrator Thomas Paine, Press Secretary Ron Ziegler, astronaut Frank Borman, and aide Bob Haldeman.

Air Force One arrived on Johnston Island at five p.m. on July 23. The Presidential entourage was greeted by a few military dignitaries and then walked across the tarmac to their waiting helicopters. The President took time to shake hands with both military and contractor personnel sta-

Fig 9D (above)—This photo shows *Hornet's* C-1A COD aircraft taking off with Admiral John S. McCain (CINCPAC) en route to the aircraft carrier.

Fig 9E (left)—President Nixon's helicopter flies low over the USS *Carpenter* in salute to the ship's officers and crew for providing support to their over-water flight.

Fig 9F—*Arlington's* CO, Captain Hugh Murphree, greets President Nixon as he steps off *Marine One* onto the ship's flight and antenna deck. Next to Nixon is Secretary of State William Rogers. Behind him are National Security Adviser Henry Kissinger and astronaut Frank Borman.

tioned on the remote base. Just a few minutes later, the party took off in two HMX-1 "white top" helicopters, trailed by a third cargo chopper.

Johnston Island was also used as a stop-over point for other dignitaries travelling to or from the *Apollo 11* recovery event. Admiral John S. McCain (Commander in Chief Pacific) flew to Johnston Island on July 23 to greet *Air Force One* when it arrived later in the day. McCain and other high ranking naval officers then flew on to the *Hornet* aboard its COD aircraft to spend the night on the ship.

The USS *Carpenter* (DD-825) was tasked as a navigational aid and plane guard, in case one of the HMX-1 helicopters needed to make an emergency landing. It was positioned 100 miles south of Johnston Island about halfway between the atoll and *Arlington*. The HMX-1 helicopters flew right over it on their way to and from *Arlington*.

As it turned out, bad weather in the primary landing area forced the *Apollo 11* landing area to be moved 240 miles downrange (north-northeast) and much closer to Johnston Island, greatly reducing the over-water flight risks.

At 1738 local time, just eighty-minutes after leaving Johnston Island, the President's group landed on the major communications relay ship *Arlington*. Members of the ship's crew formed an honor guard to "pipe" the President onboard, a traditional Navy welcoming protocol. Several hundred others ringed the flight deck in a hollow-square formation. The Commanding Officer, Captain Hugh Murphree, greeted President Nixon as he stepped onto the deck.

The President made a brief speech to the crew, saying, "It is a great privilege be on this ship and to participate in this historic occasion," referring to the *Apollo 11* splashdown. He recalled his World War II naval service as an air operations officer. He mentioned that while he had not personally been stationed aboard an aircraft carrier during WWII, Secretary of State Rogers had.

Frank Borman then spoke, reminiscing about his *Apollo 8* mission six months earlier, which *Arlington* supported as well.

Nixon went to his cabin to freshen up for a few minutes and was described by the press as being in very good humor for the rest of the evening. The presidential party toured the ship and chatted with sailors in the main enlisted men's mess hall. He even autographed the sling of an apprentice fireman who had recently suffered a steam-burned arm, and joked about trying not to get seasick in the rolling seas. Nixon expressed confidence that the lunar landing mission would have a safe ending. The All-Star baseball game, which had taken place that day and was won by the National League, was also a common topic of conversation. Nixon retired to his own cabin for a private dinner and early bed time due to the 0400 wake up call.

Fig 9G—With most of the ship's crew on the antenna deck, the President and his party performed the inspection of an honor guard, although most of the activity consisted of shaking hands and exchanging personal greetings.

On the eve of your epic mission, I want you to know that my hopes and my prayers—and those of all Americans—go with you. Years of study and planning and experiment and hard work on the part of thousands have led to this unique moment in the story of mankind; it is now your moment and from the depths of your minds and hearts and spirits will come the triumph all men will share. I look forward to greeting you on your return. Until then, know that all that is best in the spirit of mankind will be with you during your mission and when you return to Earth.

President Richard Nixon's telegram to the Apollo 11 *astronauts before launch.*

President Richard M. Nixon on the flight deck of *Hornet* greeting sailors as he walked to *Marine One* to head back to Johnston Island.

Fig 10A—The *Apollo 11* Saturn V space vehicle lifted off with astronauts Neil Armstrong, Michael Collins and Buzz Aldrin, Jr., at 9:32 a.m. EDT July 16, 1969, from Kennedy Space Center's Launch Complex 39A.

Fig 10B—The insignia of *Apollo 11*.

Chapter Ten
Flight of *Apollo 11*

The May flight of *Apollo 10* was a successful dress rehearsal for the first lunar landing mission and Pacific Ocean recovery process. All launch and spacecraft systems had performed remarkably well during its eight-day flight, setting the stage for a first-ever lunar landing only two months later.

In the early morning hours of July 16, 1969, Buzz Aldrin, Neil Armstrong and Michael Collins were secured into the *Apollo 11* command module *Columbia* on the launch pad at Kennedy Space Center. ARRS rescue helicopters hovered offshore at the launch abort site and the USNS *Vanguard* (T-AGM-19) was on-station 1,000 miles downrange. Half a world away, the range tracking ships USNS *Mercury* (T-AGM-21), USNS *Redstone* (T-AGM-20), and USNS *Huntsville* (T-AGM-7) were stationed in the southwest Pacific to relay communications and collect instrumentation data associated with the Trans Lunar Injection burn.

Just below the equator and about 400 miles southwest of Christmas Island (part of the Line Islands chain), *Hornet* patiently patrolled her assigned launch-abort area in the dark morning hours. Had the flight required an early termination while in Earth orbit, *Hornet* was positioned to make the recovery. This location also afforded *Hornet* the ability to handle problems that might arise with the TLI burn procedure on the second orbit.

At 9:32 a.m. EDT, the Saturn V engines were ignited, hurling the *Apollo 11* "launch stack" and it's occupants eastward. Eleven minutes later, they were in Earth orbit. During its only full orbit, the Apollo crew stowed their launch-related equipment (helmets, etc) and prepared the spacecraft for its 239,000 mile, three-and-one-half day journey to the Moon.

At a mission elapsed time of two hours and forty-four minutes, the spacecraft was orbiting 106 miles above the Earth and had completed roughly one-and-one-half orbits. After passing over the Solomon Islands, the S-IVB booster engine was fired (a command

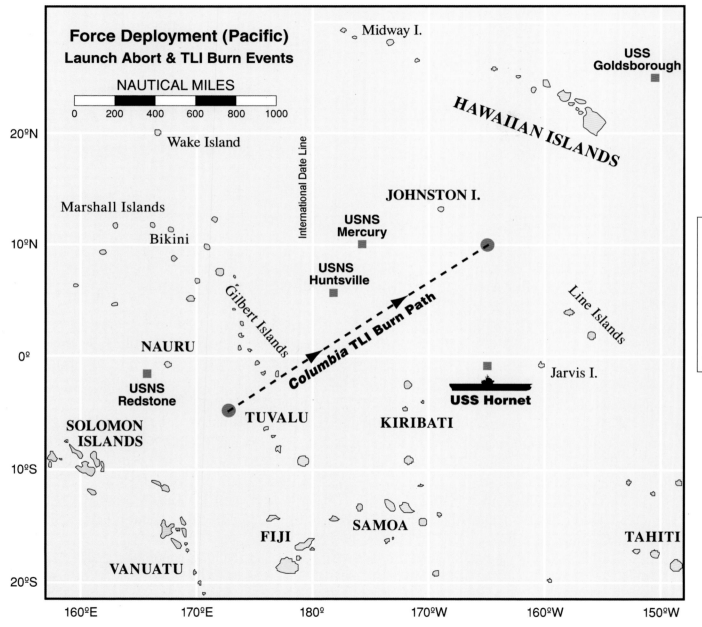

Fig 10C—Derived from the TF-140 Press Kit, this map shows the relative positions of the instrumentation ships and recovery forces in relation to the path taken by the *Apollo 11* spacecraft during its TLI burn.

relayed by the USNS *Mercury*), propelling the spacecraft on a course for the Moon. Even though it was early morning on *Hornet*, NASA's John Stonesifer, the director of quarantine operations for this mission, led several members of the recovery team up to *Hornet's* flight deck. They witnessed a small flicker of flame in the tropical sky above as the rocket boosted *Columbia* out of Earth orbit.

Once the spacecraft changed course, *Goldsborough* was temporarily released from TF-130 to continue normal naval operations until the landing operation began eight days later. *Hornet* leisurely sailed a northwesterly course along the Mid-Pacific Line "deep space abort line" for hundreds of miles, angling roughly from 165° west longitude on the equator (0° latitude) to 175° west longitude at 15° north latitude. She had to be in a very specific area each day, prepared to handle any contingency that might cause an abort situation during the long flight to or from the Moon. However, she also had to end up in position at the planned End-Of-Mission location for the recovery of a successful mission.

The Apollo spacecraft's maneuvering "footprint" during its re-entry event extended about 1,200 to 2,070 miles from where it entered the Earth's atmosphere. The design of the Command Module created a moderate lifting-body effect. NASA's initial plan called for

John Stonesifer remembers: "It was pitch black when we received word about the upcoming TLI burn. I led a few of the NASA landlubbers from the wardroom to the flight deck via some of the outboard ladders and flight-deck edge catwalks—the kind of steel grating through which you can see the water below. Once on deck, believe it or not, I put my hands out front to feel my way among the aircraft. The next day I led Ben James, the NASA Public Affairs officer, along the same route during daylight. When Ben looked down and saw the water he said, "John, if you ever take me this way again I will throw you overboard."

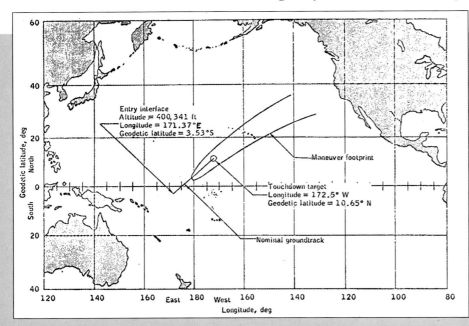

Fig 10D—This chart from NASA's *Apollo 11* Press Kit shows the nominal, or planned, re-entry point of the CM, its optimal flight path through Earth's atmosphere and the footprint (or corridor) within which it could be maneuvered to a successful landing.

Fig 10E—The ship's massive B&A crane was expected to retrieve the CM from the water, however, the crew also practiced using a Tilley crane as an emergency backup. Shown here lifting the boilerplate out of the water, this mobile crane was normally used for removing crashed aircraft from the flight deck.

the flight control system to "fly" the CM 1,475 miles downrange from the re-entry point, heading in a northeasterly direction. The re-entry control program would "dip" the spacecraft occasionally to gradually bleed off speed and lose altitude without an excessive G-load (i.e., gravity pull) on the astronauts until the parachutes could successfully deploy. *Hornet* slowly headed toward this location.

On July 20, Armstrong and Aldrin entered the Lunar Module *Eagle* and descended to the lunar surface, leaving Michael Collins alone in the Command Service Module orbiting the Moon. Several hours later, Armstrong stepped onto the Moon's surface stating, "That's one small step for a man, one giant leap for mankind." Aldrin descended shortly thereafter and they proceeded with their lunar surface activities. The astronauts deployed a variety of scientific instruments, chatted with President Nixon by "telephone" and collected forty-six pounds of lunar rock and soil samples. Much of this activity was eagerly watched after-the-fact on *Hornet's* closed-circuit TV. After spending twenty-one-and-a-half hours on the surface, the *Eagle* lifted off and docked with the CSM. The next day, the LM was jettisoned into lunar orbit and left to its fate while *Columbia* began its journey back to Earth.

While the outside world breathlessly watched these *Apollo 11* activities, life aboard *Hornet* went on, albeit with more urgency than before. All hands knew they were about to have the world's spotlight shining on them. NASA flight surgeon Dr. William Carpentier and MQF Technician John Hirasaki isolated themselves from the rest of the crew by locking themselves inside the MQF. This was a precaution to ensure they didn't introduce an Earth pathogen, such as the common cold, into the quarantine process.

SIMEX training for the recovery operation continued day and night to meet the requirement for training three men to be capable of handling each key recovery activity. The Captain also held "Presidential rehearsals," practicing the public ceremonies that would follow a successful recovery. Each ceremony was rehearsed with the color guard marching, band playing and the press snapping photos. Schedules were planned to the minute. There were stand-ins for the astronauts and other dignitaries. *Hornet's* Chaplain, Lieutenant

Commander John Piirto, ably played the role of the President Nixon. To avoid conflict, yet ensure all the members of the media captured their stories, members of the press were assigned specific locations on recovery day. The Secret Service worked with the ship's Master-at-Arms security personnel and Marine detachment to create a comprehensive security plan for the President's visit. The blueprint called for many passageways and ladders to be secured from any human access during the time the President was onboard.

The many Navy, media and White House communications systems were constantly tested. During the two press conferences each day, the media was kept informed about the status of the spaceflight in addition to interviewing key recovery personnel. Special menus were created for the occasion, the mess decks remained open almost constantly, final personnel assignments were made and a newsletter was printed specifically for the period when the astronauts were onboard.

On July 22, *Hornet* came alongside USS *Hassayampa* and the two ships performed an underway replenishment of fuel and supplies. *Hornet* was provided whatever supplies she needed to accomplish her mission, even if an extended stay at sea became necessary due to unforeseen quarantine issues.

By Wednesday, July 23, *Arlington* was steaming to its assigned location 250 miles northeast of the planned recovery area, making preparations to receive President Nixon and his party for their overnight

Fig 10F—An excellent overhead photo taken of *Hornet* and *Hassayampa* performing an underway replenishment two days before splashdown. Note the 4 E-1B Tracers lined up forward of the *Hornet's* island and one SH-3D Sea King and the C-1A Trader "COD" behind it.

stay. *Carpenter* was on station about halfway between Honolulu and *Arlington*, to act as a navigation aid and potential rescue ship for the Presidential HMX-1 helicopters. *Hornet* was on station at the planned End-Of-Mission landing area roughly 1,200 miles southwest of Honolulu, (450 miles south of Johnston Island). This location (latitude 10-56° north, longitude 172-24° west) was based on precise calculations by NASA's computers about the optimum "re-entry window" for a spacecraft returning from the Moon on that particular day. The spacecraft was programmed to fly 1,475 miles (1,285 nautical miles) from its atmospheric entry point.

To maintain its position in the expansive ocean, *Hornet* used its sonar to chart four undersea mountains and plot the exact splashdown coordinates accordingly. Captain Seiberlich named these peaks *Apollo 11*, *Aldrin*, *Armstrong* and *Collins* after the space mission. During the day, twenty-five additional newsmen from the White House Press Corps covering President Nixon's worldwide tour arrived via COD flights from Johnston Island.

Later, however, the weather situation worsened as a major storm system moved into the area. Thunderstorms and high winds aloft risked damaging or collapsing the spacecraft parachutes and the rough seas made recovering the astronauts and Command Module dangerous. *Hornet* relayed the on-scene weather information to Johnson Space Center, while a Navy WC-121N reconnaissance aircraft based in Guam surveyed the larger frontal area. NASA had already

Fig 10G—*Hornet* steaming toward the initial splashdown location in the Pacific. The E-1B Tracer and C-1A Trader aircraft are spotted on the flight deck near the bow with Helicopter #66 just to the side of the island superstructure.

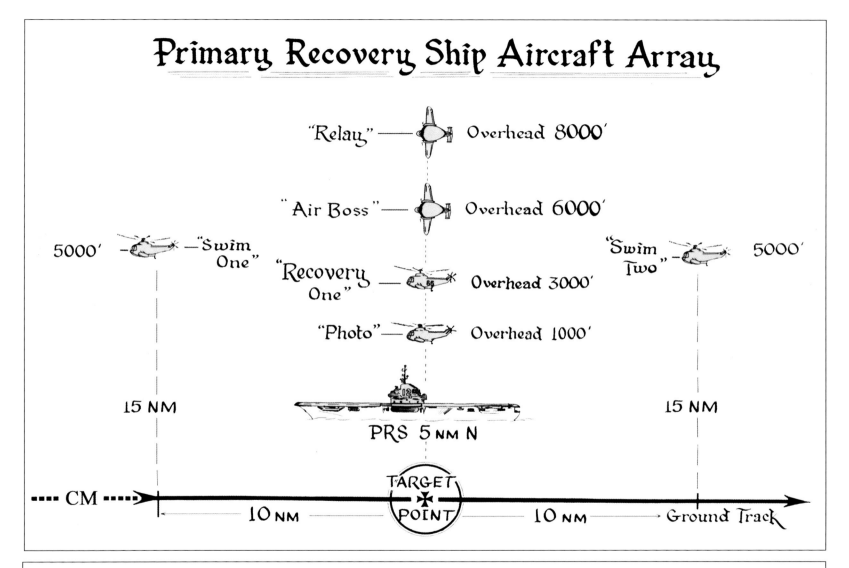

Fig 10H—From the *Hornet Cruise Report*, this chart shows the locations of the tracking and recovery aircraft were carefully planned, based upon the re-entry corridor, or ground track, of the CM. Not shown are the ARRS contingency aircraft stationed 190 miles up-range and down-range from the ship's airgroup.

received a warning from a national military weather operations group, whose data was based on the then-classified Defense Meteorological Satellite Program (DMSP) satellite system. For safety reasons, Mission Control decided to move the splashdown point northeast, closer to Hawaii and Johnston Island.

Fourteen hours before splashdown, the Capsule Communicator (CapCom) in Houston radioed the incoming spacecraft that bad weather was "clobbering" the targeted landing point. CapCom informed them: "The new aiming point is 1,500 nautical miles down range to guarantee uplift control. The weather in that area is super, with scattered clouds, ten-mile visibility and six-foot seas. *Hornet* is sitting in a great position to get to that new location."

In reality, this last minute change of 240 miles (215 nautical miles) required *Hornet* to steam at high speed to ensure its arrival at the appropriate time. Since the computerized TRANSIT satellite navigational aid had not been installed, the overcast skies during this passage made the navigation aspects difficult. Throughout the night, the navigation crew relied on dead reckoning techniques to maintain the proper course and speed to ensure the ship would arrive at the new EOM location on schedule. Finally, a short time before *Apollo 11's* re-entry, Commander Chris Lamb and Quartermaster 1st Class Howard Mooney were able to sight stars through breaks in the overcast and obtain a celestial fix of the ship's position using standard Navy sextants. *Hornet* was only a few miles from its intended position, which was twelve miles upwind of the splashdown point.

Although *Arlington* was not assigned to be in the primary recovery area, the change in the splashdown point resulted in her crew having a front row seat for the recovery operation.

Since there was no time left for one last practice, it was fortuitous that sixteen full SIMEXs had been completed, lessons learned and all crew assignments set (*see* Appendix D). Due to the Moon germ issue, *Apollo 11* required a major alteration to the flight plan used for previous recoveries. The astronaut retrieval helicopter became an integral component of the quarantine process, carrying the decontamination swimmer and his equipment to the splashdown scene and then ferrying the astronauts back to the ship. This required the primary and secondary UDT swim teams to be carried in other helicopters. Additionally, night SIMEXs had revealed a predisposition of the helicopters to converge on the area of

Quartermaster 3rd Class Rolf Sabye, a member of the plotting team in the chart room that morning, mused about how historically appropriate it was that *Hornet* used heavenly bodies, and an ancient mariner technique, to position herself to recover the first humans who walked on another heavenly body.

the expected Command Module landing—a hazardous situation for two reasons: One was the possibility of a mid-air collision in the dark skies over the featureless ocean. The other involved the CM's apex cover and drogue parachute assemblies, which fall out of the sky a few minutes behind the Command Module but in close physical proximity.

The final aircraft operations plan called for four helicopters (SH-3 Sea Kings) to deploy from *Hornet*, while a fifth one was warmed up and ready to go on the flight deck. The plan required all helicopters to be pre-positioned relative to the target point well before the splashdown occurred, and to maintain that location until released. Each aircraft was assigned specific activities, which varied depending on the stage of the recovery operation, and had a unique radio call-sign for instant identification:

- *Swim One* was a UDT swimmer delivery platform. It carried a normal aircrew of four plus a three-man UDT team with recovery equipment.
- *Swim Two* was a UDT swimmer delivery platform. It carried a normal aircrew of four plus a three-man UDT team with recovery equipment.
- *Photo* served as a photographic platform. It carried two pilots, one aircrewman, one Navy still photographer and one NASA videographer.
- *Recovery One* was the primary astronaut retrieval platform. It carried a normal aircrew of four plus a NASA flight surgeon and the UDT decontamination swimmer, in addition to the BIG suits and decontamination equipment. Prior to hoisting the astronauts up, the pilot and copilot would put on oxygen masks. One aircrewman would operate the hoist system and give directional inputs to the pilot. The second aircrewman would assist the astronauts in transferring from the Billy Pugh rescue net into the helicopter's cargo compartment. Both aircrewmen would don respirators and protective gear for their hands and feet. *Recovery One* would return to the PRS carrying the three astronauts but not the UDT "decon" swimmer, who stayed with the floating CM.
- *Swim Three* was an emergency backup that remain in standby status on the flight deck of the PRS. It carried two pilots, two aircrewmen and one UDT swimmer.

The plan called for three VAW-111 airborne early warning aircraft to orbit above. *Air Boss* was assigned as the on-scene commander of all aircraft until *Hornet* arrived at the splashdown point. Another, call sign *Relay,* was assigned to relay the on-scene UHF and VHF communications among the recovery units back to TF-130 in Hawaii. The third was kept as a backup in case of failure of either of the first two.

The planning and training was now done—all that was left was to perform the recovery operation while the President of the United States and 500 million people watched.

Fig 11A—UDT Swimmers prepare for *Hornet's* retrieval of *Columbia*.

Chapter Eleven
Splashdown and Recovery

Before dawn on July 24, the key elements of this epochal event were inexorably making their way toward *Hornet*. *Columbia* was about to plunge into the Earth's atmosphere, the President's *Marine One* helicopter was preparing for the short flight in, and a large percentage of the world's population were turning on their television sets. Well before noon, one of mankind's most momentous events would be added to the history books. In the interest of readability, event times of this day are given in local military time as taken from *Hornet's* logbook (seven hours behind U.S. EDT and eleven hours behind GMT).

Few members of the recovery teams were able to catch much sleep. The ship steamed at a high speed all night, churning through the ocean swells, while the tropic air in the berthing areas was warm and humid. Everyone had to rise at "zero-dark-hundred hours" to ensure they were in their assigned positions well before the day's events took place. The USAF Apollo Range Instrumentation Aircraft (ARIA) and Airborne Lightweight Optical Tracking System (ALOTS) units had to be in position along the 1,500-mile re-entry corridor one-half hour before the expected arrival of the spacecraft. The ALOTS EC-135N left Wake Island at 0224 to orbit above the Gilbert Islands, in position to film the atmospheric re-entry of *Columbia* and the resulting fireball. Further downrange, other ARIA EC-135N tracking and communications relay aircraft were at their assigned positions. The two Aerospace Rescue and Recovery Service (ARRS) contingency recovery aircraft, call signs *Hawaii Rescue One* and *Hawaii Rescue Two*, reported being on-station at 0504.

At 0420, while still twenty miles from the splashdown point, *Hornet* launched its airgroup to assume their patrolling areas. *Air Boss*, the E-1B Tracer carrying *Hornet's* director of aircraft operations, Commander George Patch, orbited over the ship at 6,000 feet, in position to monitor and coordinate air traffic in the immediate recovery area. Another E-1B, call sign *Relay*, orbited over the recovery area at 8,000 feet and established radio links from the recovery area to the headquarters of TF-130 in Hawaii.

Recovery One (Helicopter #66) took off with its four-man crew, plus the UDT decon swimmer Hatleberg and NASA flight surgeon Dr. Carpentier. It remained near *Hornet* until after the splashdown location was identified. NASA photographer Lee Jones, operating a 16mm motion picture camera, and HS-4 photographer Milt Putnam, taking still photos with various 35mm cameras, boarded *Photo* (#70) for their flight. The crews of *Swim One* (#53) and *Swim Two* (#64) loaded their UDT teams and began patrolling their assigned locations, twelve miles uprange and downrange from the ship.

Once the recovery aircraft were a safe distance away, *Nighthawk 2* and *Nighthawk 3*, carrying the President's staff, Secret Service agents and other personnel, were launched from *Arlington* and landed on *Hornet* eighteen minutes later. At 0512, *Marine One*, flown by Lieutenant Colonel Edward Sample with copilot Major Joe Moody, settled onto *Hornet's* flight deck. President Nixon was announced aboard in formal Navy tradition. Using the ship-wide intercom, Quartermaster 1st Class Howard Mooney tapped eight times on a brass shell casing (a procedure known as known as *bonging*) and announced to all hands: "United States Arriving, Flight Deck." Moments later, the President and his entourage emerged from the helicopter. They walked through a small honor guard of crewmen and were greeted by Admiral John S. McCain (Commander-in-Chief Pacific), Rear Admiral Don Davis (Commander Task Force 130), Captain Carl Seiberlich and Dr. Thomas Paine, the Administrator of NASA.

The President was escorted down the ship's escalator into Hangar Bay #2 while the HMX-1 helicopters were towed to the back of the flight deck and tied down. These three aircraft were maintained in "ready to fly" status, but kept out of the way of the recovery activities. Nixon met with several NASA and Navy officials, who briefed him on the recovery schedule and welcoming ceremony that would be televised worldwide. John Stonesifer, the NASA official responsible for overseeing the quarantine procedures, gave the President an

Fig 11B—The President's helicopter landed on the flight deck while the sky was still dark. Colorfully clad flight deck crewmen acted as "rainbow sideboys" to help welcome the President aboard.

John Stonesifer recalls: I enjoyed briefing President Nixon and Admiral McCain about the importance of the quarantine program and the precautions we must take. At one point in the conversation, I explained that two NASA folks had volunteered to enter quarantine with the astronauts, one a doctor to exam the astronauts and administer care if required, and an MQF engineer to keep all the systems operating until they got back to Houston. President Nixon put his hands on his hips and said, "When I was in the Navy I never volunteered for anything." We all had a great chuckle!

overview of the Biological Isolation Garment (BIG suit) and Mobile Quarantine Facility (MQF). He also explained the quarantine program that would be enforced once the spacecraft landed.

While not highly publicized, NASA had already decided that if the quarantine process was compromised in some manner, and a risk of infection to others was deemed probable, the entire ship and its crew would remain at sea for twenty-one days. President Nixon and his staff, of course, could not afford to take a three-week vacation, floating around the Pacific Ocean. Stonesifer had worked out a procedure with the Secret Service to alert them if a problem arose *before* the President was exposed to potential Moon germs, allowing them to immediately whisk him off the ship in advance of the astronauts' arrival.

The President and his entourage then went up to the Flag Bridge, where they were served a small breakfast and provided an overview of the upcoming events. *Hornet* steamed slowly north to remain upwind of the splashdown point, while still ensuring the recovery

Fig 11C—With Admiral McCain looking on, NASA quarantine manager John Stonesifer briefs President Nixon on the quarantine process and its major components, such as the MQF and BIG suits.

aircraft were perfectly positioned. After this brief flurry of aircraft activity was over, an air of quiet expectation fell over the 2,000 people onboard as they waited to fulfill the final stage of President Kennedy's challenge from 1961. President Nixon seemed to genuinely enjoy this period of calm, quietly chatting with Admiral McCain as well as senior Navy and NASA staff. At one point, he and some of *Hornet's* crew forty feet below on the flight deck exchanged light-hearted remarks.

Thirty minutes prior to re-entering the Earth's atmosphere, *Apollo 11's* Command Module separated from the Service Module. The automated entry profile was designed to ensure that *Columbia* hit the Earth's atmosphere at a six-and-one-half degree angle for successful atmospheric re-entry and a safe landing. If it hit at too slight an angle, it would skip off into outer space. If it hit at too steep an angle, it would be going too fast and could burn up. Since communications with the spacecraft went silent during this period, neither NASA nor the recovery team knew for certain what was happening.

Fig 11D—With the sun just beginning to rise, President Nixon enjoyed some banter with the flight deck crew. Frank Borman is waving to a few of the personnel stationed in front of the superstructure.

The ALOTS aircraft was positioned at 41,000 feet, heading directly toward the expected point in the sky. At 0528, they spotted the spacecraft beginning its atmospheric re-entry just southeast of Abemama Atoll, one of the Gilbert Islands. They filmed the growing fireball as it sped past on a northeasterly course toward the End-Of-Mission target point. With the camera picking up many of the re-entry components, they banked sharply to keep the camera lens pointed toward the spacecraft until it went out of sight into the rising sun. They shot 108 seconds of film and, with their mission accomplished, headed for Hawaii.

Moments later, *ARIA 3* (the uprange ARIA) reported a visual contact, but was unable to raise the astronauts over the radio. The communications blackout period lasted three minutes and thirty-four seconds, after which, a brief contact with *Columbia* was established

via *ARIA 4*. Shortly thereafter, the two ARRS HC-130Hs reported contact on the S-band radio channel.

At 0540, *Hornet's* airborne lookouts spotted the fireball to the southwest as *Columbia* burned through the early-dawn sky, leaving a 200-mile trail of flame. Because of the scattered cloud cover, it was impossible to see the Command Module from *Hornet's* bridge. When it was 150 miles away, the ship's primary air search radar began tracking the speeding spacecraft. Moments later, a double sonic boom rolled across the otherwise silent flight deck, indicating it had slowed down enough for the drogue parachutes to begin opening at 24,000 feet. Finally, *Hornet* was able to establish permanent voice contact with the spacecraft. The astronauts reported they were in good shape.

Shortly thereafter, when *Columbia* had descended to 10,000 feet, the three huge main parachutes blossomed out. Almost immediately, *Air Boss* CDR George Patch radioed he had gained visual contact. The spacecraft's thirty-two-feet-per-second descent was sporadically hidden from the air units by overcast skies and only intermittently visible to a few

Major Charles Hinton, the ALOTS aircraft navigator, recalls: "Everyone in the aircrew was a bit surprised by the fireball track as the CM re-entered Earth's atmosphere. Even though we were orbiting at 41,000 feet, we somehow felt the spacecraft should come downwards from someplace "up there." It actually appeared to come in from below (two degrees below the horizon) and then passed overhead into the dawn sky to the east."

Fig 11E (opposite)—Traveling at 24,700 mph, *Columbia* began its fiery plunge 400,000 feet above sea level over the Gilbert Islands, heading in the general direction of Hawaii. This NASA drawing depicts the early moments of the communications blackout.

Fig 11F (right)—Since it was fairly dark, with cloudy skies, no one clearly observed the deployment of the drogue, pilot or main parachutes. This is a NASA artist's drawing of the deployment of the main parachutes after the spacecraft had been slowed by the drogue parachutes.

personnel on *Hornet*. The pilot of *Swim One* sighted the flashing beacon as the spacecraft passed through 2,000 feet, maintained visual contact during the final minute of flight and announced its landing over the radio.

Splashdown occurred at 0550 local time on July 24. Impact on the ocean was one-and-one-half miles from the target point and about thirteen miles from *Hornet*. It was six miles from *Arlington;* a few members of her crew saw the spacecraft just before it hit the water. The splashdown point (latitude 13.19° north, longitude 169.09° west) was 240 miles south of Johnston Island and 920 miles southwest of Pearl Harbor. *Apollo 11* logged 950,000 total miles traveled during its 195 hours and 18 minutes of flight.

The TV broadcasters complained a little about not being able to film the actual moment of impact. To their dismay, Captain Seiberlich positioned the ship upwind at the outer limits for visual observation of the splashdown. NASA had given him a ten nautical mile diameter recovery zone and their protocol only required the recovery helicopters to be at the spacecraft landing site, not the ship. While under parachute descent, the Command Module vented excess propellants that were highly toxic. The spacecraft's internal cabin pressure was equal to an altitude of 40,000 feet, so the risk of also venting lunar dust was minute, but not impossible. The Captain did not want to risk endangering the presidential party nor the ship's crew and maintained a large buffer zone until the spacecraft was in the water.

The Command Module came to rest in a Stable 2 position (upside down) due to eighteen-knot wind gusts filling the parachutes and dragging the spacecraft onto its side after landing. Green dye marker was released into the ocean to enable aircraft to locate the CM more effectively. Moments after impact, how-

Fig 11G (left)—After the CM splashed down, the first two helicopters on the scene were *Swim One* (shown) and *Photo*, from which this picture was taken. The uprighting bags are almost fully inflated and well along in turning the spacecraft apex-up. A main parachute is floating in the water to the right.

Fig 11H (opposite)—The other UDT divers having completed their assignments, Hatleberg prepares to jump from Helicopter #66. He can be seen standing in the cargo hatch timing his leap from the helicopter.

Hatleberg remembers: "Now you couldn't talk understandably through the BIG suit's facemask filters, plus there was the helicopter rotor-blade noise, which drowned out all but a shout. So, I didn't think anybody was going to say anything and I decided to focus on just following the procedures to maintain the quarantine.

"Buzz Aldrin was the first astronaut to exit out of the hatch. He shook my hand and stated something like, "Garble-garble-garble." As a good Navy-trained man who always repeats back an instruction, I replied, "Garble-garble-garble" in response. Now, Buzz had probably said some very meaningful words, but they just sounded like garble to me, so I actually said the words "garble" back, not knowing what else to say."

Thus transpired the first exchange of words between Spaceman and Earthling!

ever, two helicopters were hovering overhead, even though the astronauts were unable to communicate with the outside world since the VHF antennas were underwater. The spacecraft's crew triggered the release and inflation process of the three uprighting bags. While the astronauts hung by straps from their couches, the white bags slowly but methodically inflated, and seven minutes later, *Columbia* was in Stable 1, floating apex-up with its side exit hatch out of the water.

Although the secondary UDT team was in *Swim One*, the nearest helicopter, the brief righting period allowed the more experienced primary team to arrive on-scene. At 0558, *Swim Two* made its initial pass by the spacecraft, allowing Seaman John Wolfram to jump from about ten feet above the water. Wolfram swam over to the CM and peered in the window to check on the condition of the astronauts. After receiving a thumbs-up from Buzz Aldrin, he attached and deployed the sea anchor, an underwater parachute designed to slow the drift rate of the spacecraft in the choppy seas.

Swim Two then made a second pass, allowing two more members of the swim team, LTJG Wes Chesser and Quartermaster 3rd Class Michael Mallory, to jump in. The crew chief of *Swim Two* shoved the flotation collar package out of the Sea King's hatch. The frogmen quickly inflated the flotation collar using two air bottles and then connected it to the wildly bobbing spacecraft by using bungee lines. Once the Command Module was stabilized, *Swim Two* dropped the two seven-man life rafts, which were quickly inflated and deployed. One was tethered directly in front of the CM hatch as a work platform for biological decontamination and helicopter hoist operations. The other was tied so it would remain a distance of 100 feet upwind. This allowed the UDT swimmers, who were using standard SCUBA equipment and wet suits rather than BIG suits, to maintain an emergency response position without fear of biological contamination.

The initial UDT swimmers completed their CM stabilization activities within several minutes, preparing the scene for the next phase, the crew and spacecraft decontamination process. *Swim Two* departed the area, enabling *Recovery One* to make its first pass by the spacecraft. Lieutenant Hatleberg, the decon specialist, jumped into the water wearing a swim suit, face mask and fins. During a second pass, aircrewmen lowered the decontamination equipment to the UDT swimmers and it was placed in the decon raft tethered to the front of the Command Module. This consisted of a bag containing four BIG suits (three for the astronauts and one for Hatleberg) and two decontamination solution containers. Hatleberg donned his BIG suit before signaling the astronauts to carefully open the CM hatch. He handed in a bag containing their three garments and closed the hatch.

Within a few minutes, the astronauts donned their BIGs, the hatch was reopened and Hatleberg assisted them out into the decon raft. As each one crossed the threshold of the hatch, they inflated their water wings (life vest) over their BIG suits.

It required three attempts by Hatleberg to close the hatch, because of a stuck handle. Michael Collins recycled the locking mechanism of the hatch and they slammed it shut. Hatleberg sprayed the hatch and vents with betadine to kill any potential lunar pathogens, making sure to stay clear of hazards posed by the spacecraft's maneuvering thruster chemicals.

Following the NASA protocol, all four people then scrubbed each other with sodium hypochlorite to ensure no pathogens could be carried on the outside of their BIG suits. During this period of time, the facemasks of the astro-

Fig 11I—UDT decon swimmer Hatleberg is wiping down one of the astronauts, using a cloth mitt on his left hand that was sprayed with bleach. After the three astronauts were decontaminated, one of them wiped Hatleberg to ensure all potential Moon germs were dealt with.

nauts started fogging up, impairing their vision. At 0642, NASA flight surgeon Dr. Carpentier, who observed these decontamination activities closely from *Recovery One*, reported to *Hornet* that proper protocol had been followed and there were no breaks in the process. Unknown to the ship's crew and the media, this was a prearranged signal to Captain Seiberlich about whether or not to have President Nixon immediately evacuated from the ship. Based on Dr. Carpentier's observations, any potential Moon germ threat during this phase of the recovery process had been mitigated. Had the protocol been compromised, the entire ship might have been quarantined at sea for twenty-one days.

Upon receiving a hand signal from Hatleberg, Commander Don Jones maneuvered

Fig 11J—The first astronaut is being hoisted up to the helicopter in the Billy Pugh net, and can be seen about ten feet directly above the raft. The UDT swimmers are positioned near the CM in case a wave flipped over the raft or other emergency situation arose.

Recovery One into the retrieval position forty feet above and slightly to the side of the raft and Command Module. In addition to their normal flight suits, the two aircrewmen in the cargo bay wore respirators over their faces, gloves on their hands and covers on their boots to maintain the quarantine. Now that the astronauts and CM had been decontaminated, the other UDT swimmers swam closer to the decon raft to provide rescue support during the winch operation, since Hatleberg couldn't swim in his BIG suit.

Because of the facemask on the suits and the bobbing of the raft, it was hard to tell the astronauts apart. Hatleberg helped the first astronaut, believed to be Neil Armstrong, into the Billy Pugh net for the twenty-second ride up to the open hatch of the Sea King. Armstrong was assisted into the cargo/passenger hold by Chief Petty Officer Norvel Wood and Chief Petty Officer Stanley Robnett. Immediately, Dr. Carpentier performed a quick visual exam of the astronaut who then sat down on a canvas "troop seat." The Billy Pugh net was lowered for the next astronaut.

The second astronaut, probably Michael Collins, was lifted up without difficulty. Shortly thereafter, Buzz Aldrin took the "HS-4 elevator" and all three were seated next to each other in the rear of the helicopter after the flight surgeon was assured they were feeling fine. Due to the restricted visibility out of their facemasks, none of the astronauts saw the greeting sign painted on the underside of the helicopter saying "Hail Columbia." The hatch on the Sea King was slammed shut in preparation for the short flight back to the ship and the two aircrewmen moved forward toward the flight deck. At this point, heat from the astronaut's bodies started building up in their BIG suits and, combined with the smell of aviation fuel, made for an uncomfortable

Fig11K—This unusual view taken from the *Recovery One* helicopter shows the astronauts, with their life vests attached, waiting in the decon raft while Hatleberg reaches out for the Billy Pugh net, the yellow object at the bottom of the photo.

situation in the humid tropic air. *Recovery One* delayed its departure for a minute to let *Swim One* get in trailing formation and escort it in case of a malfunction.

By now, *Hornet* had come to a halt only one-half mile away from the scene. President Nixon moved to the back end of the Flag Bridge to get a better view of the events happening directly off the port side of the ship. He was joined by NASA Administrator Thomas Paine and other members of his staff. Surrounded by media photographers, they watched as the *Apollo 11* crew was hoisted into the recovery helicopter and flown back to the ship.

Fig 11L (above)—President Nixon and NASA Administrator Paine watch intently as the three astronauts are hoisted into the recovery helicopter and flown to the *Hornet's* flight deck.

Fig 11M (right)—The astronaut recovery helicopter was welcomed by a small group of sailors, NASA personnel and the band. As soon as it was placed into its ship-board configuration, it was towed to a few feet forward to Elevator #2 and lowered to the hangar deck with the astronauts still inside.

The COMNAVAIRPAC Navy band played "*Columbia, Gem of the Ocean*" as Helo #66 landed on the flight deck just aft of the #2 (port side) elevator amid a crowd of media. Once its engines were shut down, the Sea King was quickly reconfigured for ship-board maneuvering. Its rotor blades were folded back and locked in place, pyrotechnics put on "safe mode," and a tow bar attached to the front. An HS-4 crewman ran out and placed a third Apollo Command Module decal on the side of the cockpit, denoting its third successful Apollo recovery. They also placed a sign on the side of the cockpit saying "Now Hornet Plus Three."

With the crew chief riding the brakes of the helicopter for safety, it was towed onto the ship's port side elevator and lowered into the hangar deck. The Sea King was then pushed by the tow vehicle tail-first until reaching a position adjacent to the primary MQF in Hangar Bay #2.

NASA Recovery Operations Director Don Stullken and Quarantine Manager John Stonesifer were so ecstatic they grabbed the portable stairway and placed it next to the Sea King's hatch, rather than wait for the ship's crew to do so. Another NASA person stood beside the steps with a fire extinguisher filled with glutaraldehyde (a colorless liquid with a pungent odor, normally used to sterilize medical equipment) to spray every area touched by the astronaut's hands and feet. The hatch was opened, and the three astronauts exited into the bright lights of television and a crowd of dignitaries, media, and sailors who gave them very enthusiastic applause.

After the astronauts stepped down from the helicopter, with Dr. Carpentier trailing close behind, they walked briskly about ten steps and entered the MQF. Michael Collins hurried in, feeling slightly nauseated by the BIG suit in the humid tropical environment. He was followed closely by Neil Armstrong and Buzz Aldrin. The door was closed, shutting in the astronauts, along with Dr. Carpentier and John Hirasaki, the MQF technician. This ten-second event took place while the press and hundreds of sailors were cordoned off only a few feet away. It stands in stark contrast to all NASA space missions that preceded it, such as *Apollo 10,* when the ship's officers and crew greeted the astronauts with handshakes and a formal welcoming ceremony on the flight deck.

The *Apollo 11* astronauts quickly shed their BIG suits, replacing them with more comfortable light blue NASA flight suits adorned with the Apollo program badge and "Hornet Plus Three" lapel buttons. Following a brief physical exam of each, they gathered in the lounge area near the main MQF entrance. In the meantime, decontamination activities outside the MQF continued in preparation for the President's imminent arrival. Two minutes after the astronauts deplaned, the gleaming white Sea King helicopter was towed out of Hangar Bay #2 and taken up to the flight deck. The steps used by the astronauts to descend from the aircraft and the deck area leading to the MQF were sprayed with glutaraldehyde. On the port side of the hangar bay, a number of chairs were set up for various dignitaries to observe the ceremony. The Navy band came down from the flight deck and set up in this same area in order to welcome the President. During this process, the aircrewmen of Helo #66 were interviewed on live TV by Dallas Townsend, who wanted to find out how the astronauts were feeling. However, the HS-4 personnel made it clear that hand signals were all that occurred since the BIG suit masks and helicopter rotor blade noise made it impossible to have a real voice conversation.

Soon, Dr. Carpentier telephoned the bridge to announce "all ready." The President and a few others came down from the flight deck on the ship's escalator, which is attached to the starboard side of the hull. The band alerted the audience to the President's arrival

Fig 11O (left)—Astronaut Michael Collins enters the MQF, followed by Armstrong, Aldrin and Dr. Carpentier who needed to perform a brief medical check and get them into flight suits before the welcoming ceremony. The plastic tunnel that will connect the CM and the MQF is already in place.

Fig 11P (below)—This is the broader perspective on the welcoming ceremony that occurred in Hangar Bay #2. The TV camera is broadcasting the conversation between the President and the astronauts while most of the sailors are screened from viewing it by the throng of media and military journalists.

Fig 11N (opposite)—The three astronauts clad in BIG suits descend from Helo #66. The first is Michael Collins, followed by Neil Armstrong and then Buzz Aldrin. John Stonesifer (in suit) applauds. Another NASA person waits with decontamination solution loaded into a fire extinguisher to spray their path.

by sounding four ruffles and flourishes when he entered Hangar Bay #2. They struck up a rousing rendition of "Hail to the Chief" as President Nixon strode through the hangar bay to the forward entrance of the MQF. He waited at the microphone while the astronauts fumbled a bit in moving the curtain from the door's window. Armstrong, Collins and Aldrin each had a handheld microphone for communicating with the President. This ceremony was broadcast live on TV and watched by an estimated 500 million people around the world.

The President made a short welcoming speech, stating this was one of the most momentous occasions in the history of mankind. Smiling often, he engaged in light-hearted banter with the *Apollo 11* crew about various subjects, such as professional baseball league standings. At one point, he mused whether the speed of space travel made a man just a bit younger than he had been before the journey. And, if that were the case, would the *Apollo 11* astronauts now be younger than Frank Borman? Michael Collins immediately quipped "We're *all* a lot younger than Frank Borman!" The President laughed and motioned Borman to come over from the dignitary area and join the conversation for a few moments. Borman mentioned how many times Collins had used the words "fantastic" and "beautiful" to describe the Moon, reflections of Borman's own feelings during his *Apollo 8* flight. The President then commented that the world was infinitely bigger now that the Moon was accessible to humans, but that the people on Earth had never felt closer together.

Fig 11Q (opposite)—President Nixon was clearly very pleased with the day's events. He was extremely glad the astronauts made it back safely and joked with them several times.

Fig 11R (right)—After *Apollo 8* astronaut Frank Borman became the subject of a Collin's quip, President Nixon motions for him to come greet the *Apollo 11* crew.

Fig 11S—Before heading to his helicopter, President Nixon grabbed Captain Seiberlich's hand and heartily congratulated him for the outstanding job performed by *Hornet's* crew.

After promoting the two military astronauts, Aldrin and Collins, President Nixon extended an invitation for them and their wives to join him at a state dinner in Los Angeles once the quarantine period ended. He asked if that was okay and Armstrong replied with a chuckle, "We'll do anything you say, Mr. President." Chaplain John Piirto provided a prayer, and the formal ceremony came to an end after the band played the National Anthem.

President Nixon and his entourage went up the escalator to the flight deck, with onlookers describing his appearance as exuberant. Accompanied by Admiral McCain and Captain Seiberlich, he slowly walked along a line of ship's crew and airgroup personnel, shaking hands and exchanging remarks with many of them, much to the concern of his Secret Service detail.

At one point, he went over to flight deck crewmen and said "Congratulations on a great job."

The sailor, an aircraft refueler, replied simply, "Yessir, we're *Hornet*."

Knowing the legacy of the seven previous *Hornet's* in American history, the President turned to Captain Seiberlich and stated, "Tell your crew that today, *Hornet* lived up to its legacy."

President Nixon walked up the steps of *Marine One* shortly after 0810, about three hours after he arrived. Giving one of his customary arm-wave salutes to the assembled group, Nixon turned and boarded the aircraft. *Marine One* took off, performed a fly-over of *Arlington* and headed back to Johnston Island with *Nighthawk 2* and *Nighthawk 3* following behind. After a short motorcade and quick lunch at the base on Johnston Island, Nixon boarded *Air Force One* and began the next leg of his worldwide trip by flying to Guam.

Immediately after the President's departure, Admiral McCain addressed the *Hornet* crew over the ship's intercom, expressing his congratulations on their accomplishment. Ten minutes later, a VR-30 aircraft carrying Admiral John McCain and Lieutenant Commander Charles Larson was connected to the catapult and launched on its flight to Johnston Island.

After most of the dignitaries departed, things calmed down a bit and the recovery team focused on the second part of its objective—recovering the Command Module with its precious Moon rocks. Once the astronauts left the scene of the bobbing *Columbia*, Hatleberg placed his BIG suit, decontamination bottles, and mitt inside the decontamination raft and then sank it. At that point, the other UDT members got onto the flotation collar and all of them began preparing the CM for its retrieval.

Prior to the welcoming ceremony in the hangar bay, a motor whaleboat with five crewmen was put in the water and sped to the location of the Command Module. The crew's mission was to radio position information and status about the CM to *Hornet's* bridge, photograph the recovery operation and assist the UDT team in any way necessary.

Charles Larson remembers the catapult launch very well. He was a U.S. Naval Academy classmate of ADM McCain's son, LCDR John McCain, who was then being held as a prisoner of war in Hanoi. For almost two years, the Admiral (as CINCPAC) had been weighed down by having to run the Navy's war effort in Vietnam, as well as deal with his son's harsh captivity. But on this day, he was in high spirits from the flawless recovery of *Apollo 11*. The seats in the passenger cabin of a C-1A Trader face to the rear of the aircraft. McCain, who was a diminutive person, belted himself into a seat across the aisle from Larson and held a customary unlit cigar in his left hand. When the ship's hydraulic catapult suddenly yanked the aircraft forward and flung it into the sky, the Admirals' legs flew straight out and he was held in place only by his seat belt. He looked over at Larson and with a big grin said, "Now *that* would be a one dollar ride at Coney Island!"

Fig 11T—Decon swimmer Hatleberg removes his BIG suit in preparation for sinking all of the decontamination equipment to the ocean floor. *Swim Two* hovers overhead to watch for sharks or respond to any unforeseen problems.

Fig 11U—*Hornet* approaches the CM on its starboard side, using single revolutions-per-minute speed adjustments. A guide was positioned in the bow to relay minute steering instructions to the Captain, who was conning the ship. The motor whaleboat is near the spacecraft to assist recovery efforts.

Captain Seiberlich personally assumed the "conn," meaning he was in charge of the ship's maneuvering, and positioned himself on a small starboard wing, just off the bridge. It was normally used for guiding the ship while doing an "UNREP" with one of the big fleet oilers. Without having the flight deck extending hundreds of feet in front, it provided good visibility forward of the ship. Quartermaster 2nd Class Ken Hoback was on the main helm, while the Officer of the Deck was positioned so that he could relay orders between the conning officer and the helmsman.

Personnel in the Combat Information Center tracked the position of the spacecraft in relation to the ship by using the MK-56 gunfire director radar on the starboard 5" anti-aircraft gun mount near the bow. A forward observer, and a talker with sound-powered headphones, was also placed at the bow of the ship on the starboard side catwalk.

Hornet's CO judiciously used a combination of the ship's four propellers, speed and rudder to execute a series of maneuvers that placed the CM on its lee, or down-wind, side. The goal was to have the huge ship block the wind, so the spacecraft couldn't drift very fast. At a range of 1,500 yards, the ship's speed was reduced to five knots. When the Command Module was 500 yards in front and 100 feet to the right of the bow, the engines were stopped. The 44,000-ton *Hornet* inched alongside the five-ton CM until it was adjacent to the bridge. Hoback steered the ship alongside the spacecraft at such a slow speed they eventually lost "steerageway" (i.e., the helmsman was unable to control the ship using rudders). Adjustments were made by the manipulation of the ship's propellers, reversing its engines ever so slightly. Seiberlich wanted the ship stopped as the Command Module came adjacent to the bridge, rather than further aft by the starboard side elevator. The UDT team was riding on the CM and in the attached raft, and the possibility remained for them to be sucked into the propellers by the slightest mistake.

Using great foresight, one of the UDT swimmers had earlier decided he wanted his mother to see him on TV. Since all three CM-preparation swimmers had the same full-body wet suits on, they decided to apply different (non-military issue) bathtub appliqués (sunflowers, etc) to obvious places on their wetsuits. As the *Hornet* got nearer, its crew enjoyed an unexpected, yet colorful, sight.

Hornet's Command Module retrieval team threw an in-haul line over to the UDT team, which reeled in the main rope that, in turn, was attached to a special NASA lifting hook assembly. The UDT swimmers attached several steadying lines to the front of the CM, allowing the ship's crew to keep it facing the ship so it would hang from the cable at the correct angle. These also prevented the spacecraft from swinging wildly or banging the side of the ship if *Hornet* was hit by a large swell.

Showing great skill, the operator of the ship's B&A crane gently plucked the spacecraft from the water during the crest of an up-swell. The operator initially lowered it onto Elevator #3 so the ship's retrieval team could undo the bungee lines that fastened the flotation collar to it. The crane then raised it back up so the spacecraft dolly could be towed underneath it. At 0855, *Columbia* was lowered down and secured to its transport dolly. Roughly three hours after splashdown, *Columbia's* mission had ended. While the exterior of the CM was scarred in places from the heat of re-entry, its remaining kapton foil covering shined like gold as it was tugged into Hangar Bay #2 and secured near the MQF.

The recovery phase of the *Apollo 11* lunar landing mission had been completed and *Hornet* set a course for Pearl Harbor.

Fig 11V—The initial shot line has been thrown to the UDT swimmers and they are waiting for the B&A crane to lower the lifting cable. Sunflower appliqués can be seen clearly on John Wolfram's wetsuit, swimmer on top of the CM. Decon swimmer Hatleberg has removed, and sunk, his BIG suit and is now in swim trunks.

Fig 11W—This photo shows the CM being raised out of the ocean. In the water, under the B&A crane, the UDT team can be seen climbing a rope net onto the ship. The green dye from the CM can be seen floating under the elevator. On the flight deck, Helicopter #66 is being decontaminated with formaldehyde gas. In front of the helicopter, an ABC-TV camera is filming the recovery of *Columbia*.

Fig 11X—The CM was lifted onto Elevator #3 and its flotation collar removed. It is being lifted again so it can be placed on its dolly for relocation next to the MQF and connected to the plastic tunnel. Sharp eyes will notice graffiti on the collar and a reduced amount of kapton foil on the CM, which became a highly prized souvenir.

Fig 11Y— Lieutenant Commander John A. Piirto, USN, Chaplain of the *Hornet*, during rehearsal for the Welcoming Ceremony.

The prayer offered by Lieutenant Commander John A. Piirto, USN, Chaplain of the *Hornet*, after President Nixon greeted the astronauts.

Lord God, our Heavenly Father, our minds are staggered and our spirits exultant with the magnitude and precision of this entire *Apollo 11* mission. We have spent the past week in communal anxiety and hope as our astronauts sped through the glories and dangers of the heavens.

As we tried to understand and analyze the scope of this achievement for human life, our reason is overwhelmed with the bounding gratitude and joy, even as we realize the increasing challenges of the future. This magnificent event illustrates anew what man can accomplish when purpose is firm and intent corporate.

A man on the Moon was promised in this decade, and though some were unconvinced, the reality is with us this morning in the persons of the astronauts: Armstrong, Aldrin, and Collins. We applaud their splendid exploits, and we pour out our thanksgiving for their safe return to us, to their families, to all mankind.

From our inmost being, we sing humble yet exuberant praise. May the great effort and commitment seen in this Project Apollo inspire our lives to move similarly in other areas of need. May we, the people, by our enthusiasm and emotion and insight move to new landings in brotherhood, human concern, and mutual respect. May our country, afire with inventive leadership and backed by a committed followership, blaze new trails into all areas of human care.

Speed our enthusiasm and bless our joy with dedicated purpose for the many needs at hand. Link us in friendship with peoples throughout the world as we strive together to better human conditions. Grant us peace beginning in our own hearts and a mind attuned with good will toward our neighbors

All this we pray as our thanksgiving rings out to Thee, in the name of our Lord, Amen.

Fig 12A—RADM Davis and CAPT Seiberlich emcee a Navy ceremony for the astronauts

Chapter Twelve
Return to Pearl Harbor

While the recovery of crew and spacecraft from the sea was complete, there was still much work to be done. Helicopter #66 was towed to a spot on the flight deck directly behind the GE/WUI satellite terminal aft of the ship's island. Two ventilation tubes were hermetically connected via the cargo hatch, and formaldehyde gas pumped through it, to decontaminate the interior from any Moon germs that may have lingered on the astronaut's BIG suits.

The remaining aircraft and helicopters still in flight landed aboard ship. The helicopters were washed down with fresh water and six of them lowered into Hangar Bay #1. The other two remained on the flight deck to act in a search and rescue role should the need arise. The E-1B Tracers were spotted (located) on the forward starboard side of the flight deck and kept clear for C-1A and US-2B Trader launches that ferried dignitaries, newsmen and Moon rocks from the ship.

A Surface-to-Air-Recovery (STAR) pickup took place at 0920, shortly after retrieval of the Command Module, to take highlight news film back to Hawaii. USAF crewmen temporarily assigned to *Hornet* placed the reels in a special container and connected it to the balloon. On the forward part of the flight deck, they filled the balloon with helium and raised it via its nylon tether.

The HC-130H then approached from the rear of the ship, snagged the rope and carried away the package, reeling it in as they turned and headed for Hawaii.

Meanwhile, in Hangar Bay #2, a plastic tunnel was put in place connecting the MQF with the CM hatch. As soon as the tunnel was secure, Hirasaki opened the spacecraft hatch and recorded the condition of its interior by taking photos and completing checklists for switch positions. The space suits, film and other items were transferred into the MQF. These were sorted out and appropriately stored in closets and bins underneath the bunks.

Fig 12B (top left)—A close view of Helicopter #66 undergoing a decontamination process using formaldehyde gas to ensure no Moon germs lingered in its cargo hold.

Fig 12C (below)—USAF personnel assigned to the recovery force fill the balloon with helium on the forward flight deck and prepare to raise it on its tether.

Fig12D (bottom left)—The HC-130H cargo aircraft lumbers over the flight deck, snagging the tether line almost dead center. The balloon can just be seen above the aircraft cockpit.

John Hirasaki remembers: "The *Apollo 11* crew and Dr. Carpentier were busy with their assigned tasks while I mainly oversaw operations unique to the MQF (power configuration changes, interior communication system setup, transfer lock operations, and photo-documentation). My duties were split between performing post-flight activities inside the Command Module and maintaining the normal operation of the MQF."

After that, he removed the two Lunar Sample Return Containers (rock boxes) the astronauts had packaged forty-six pounds of lunar samples in and prepared them for transfer to Houston. This involved double vacuum packaging each container in heat-sealed bags, prior to immersing them in the MQF Transfer Lock, so they could be passed out to the NASA technicians waiting outside. After completing the work inside the CM, Hirasaki sealed the end of the tunnel and brought it inside the MQF. Each of these vacuum-sealed packages were placed into special shipping containers so they could be taken to

Fig 12E—Once the Moon rocks were outside the MQF, NASA technicians placed the lunar samples into special biologically isolated containers for immediate shipping to the Lunar Receiving Laboratory.

Houston well in advance of the astronaut's arrival. This allowed NASA to transport the MQF and the Command Module to Houston by separate routes.

Following a prearranged protocol, one package of lunar samples was put on each of two C-1A transport airplanes. This ensured survivability of some lunar samples in case of a catapult failure or bad weather situation. The first aircraft was launched that evening and flew to Johnston Island, 180 miles away. The sample container was transferred to a USAF C-141 and, along with some NASA personnel, flown to Ellington Air Force Base near Houston. The departure from *Hornet* of the remaining samples was delayed until astronaut medical specimens were ready for transport. Early on July 25, about six-and-one-half hours after the first Trader launch, the second C-1A departed and flew directly to Hickam AFB in Hawaii, 620 miles away. That container was placed on one of the ARIA aircraft and flown back to Houston, earning that plane the nickname "The Moon Rock Express." Within forty-eight hours of splashdown, scientists at the Lunar Receiving Laboratory were conducting preliminary tests on the priceless lunar samples.

Fig 12F—A C-1A Trader, configured to carry four passengers plus cargo, is launched from catapult #2 with half of the lunar samples onboard. This aircraft flew to Johnston Island.

Dr. Paul Rambaut of NASA remembers: "We had created special meals in advance for the flight crew and the others inside the MQF. Our primary concern was to avoid introducing any noxious Earthly chemical or organism that could cause problems with the crew that might be thought to have a lunar origin."

The astronauts settled into a routine of playing cards, reading newspapers, enduring various medical tests and catching up on their sleep. The menu contained primarily precooked, frozen entrees that were sent in via the transfer lock. The meals, such as a cheese omelet for breakfast, roast beef sandwich for lunch and chicken Kiev for dinner, were then reconstituted in a then-unique counter-top microwave oven. The MQF was guarded by the ship's Marine Detachment but passers-by were occasionally able to snap a photo of the men inside, including one of Neil Armstrong playing a ukelele.

At noon on July 25, *Hornet* held its own formal welcoming ceremony, which was captured by TV crews and broadcast on the nation's evening news by all three major networks. The event began with the playing of the "Star Spangled Banner" by the Navy band and opening remarks by Captain Seiberlich. It included Buzz Aldrin, a Colonel in the

Fig 12G—In between numerous medical tests and other post-flight activities (such as readjusting to gravity), the astronauts took some time to relax and mentally decompress from their flight.

USAF, presiding over the swearing-in of several non-commissioned officers who were re-enlisting. This was followed by the traditional cake cutting ceremony, although the cake was not served at this time. The Captain presented the astronauts with *Hornet* commemorative mugs, photographs and plaques (viewed through the MQF window, of course). There was a Thanksgiving service and at 1230, the band played "Anchors Aweigh" as the Commanding Officer departed.

Some of the ship's crew had anything but a normal routine. The ship's post office had been provided only a single *Apollo 11* recovery cachet, but over 248,000 covers arrived for stamping and cancellation. This process consumed five officers and twenty enlisted men

Neil Armstrong recalls: Dr. Carpentier took good care of us, the food and drink was excellent compared to our rations during the flight. We had a great deal of work to do getting our thoughts recorded as preparation for all the post-flight debriefings for which we were obligated.

Fig 12H—During *Hornet's* welcoming ceremony, the astronauts presided over the re-enlistment ceremony for a number of Navy and Marine NCOs.

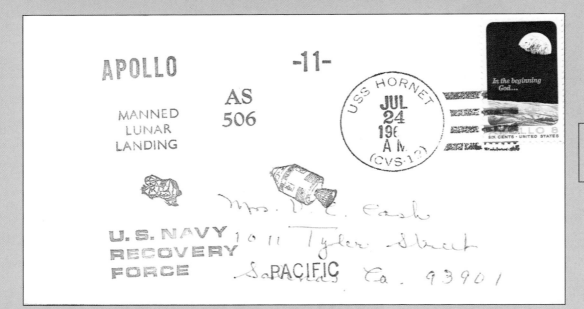

Fig 12I (left)—This is an example of the standard cachet affixed to envelopes to commemorate the splashdown and recovery of *Apollo 11*.

Fig 12J (right)—This is an example of the very scarce and much coveted "Captain's Cover" envelope design, autographed by CO Captain Seiberlich for Sgt. Joe Holt, a member of the ship's Marine Detachment.

during the fifty-two hour journey back to Pearl Harbor. The ship's store did a brisk business in official souvenir items, with baseball caps, coffee mugs and key chains at the top of the popularity list. The hottest item, however, was not for sale from the store. A lucky few were able to snatch a small piece of the kapton foil that coated the outside of the Command Module. These finger-length gold strips quickly became the "must have" souvenir of the recovery operation.

During this same period, personnel affiliated with NASA, UDT and the media, disassembled, packed and/or positioned as much equipment as they could on the flight deck and in the hangar bays to expedite the off-loading process at Pearl Harbor or later at Long Beach. ABC-TV's correspondent Keith McBee decided to fly back to Hawaii to speed up delivery of the original newsreel tapes for the entire recovery operation. The tape canisters were three inches deep, roughly twenty inches square and weighed twenty-five pounds each. There were four tapes, one pair from each of the two recording vans, but there was only room for two tapes under the seat in the COD plane. After struggling to get one set stowed away, he turned to *Hornet's* Executive Officer, Commander Chris Lamb and asked him to keep the second set in case the COD plane crashed. McBee made it to Hawaii okay and CDR Lamb maintained possession of the other tapes, never having been notified of what else to do with them.

Just prior to entering Pearl Harbor, *Columbia* was brought up to the flight deck and placed forward of the island so it could be viewed by the public. When the ship entered the main entrance channel to the historic harbor, the crew "manned the rails" in their dress white uniforms. A broom was lashed to the mast, the Navy tradition indicating "mission accomplished."

Two tugboats struggled to gently push *Hornet* next to Pier Bravo. The Navy band on *Hornet's* flight deck, standing next to the Command Module, struck up a tune. A crowd estimated at 2,500 people waited to greet the *Apollo 11* crew with a hero's welcome and a little Hawaiian pomp and circumstance—flower leis, ukelele music and hula dancers. *Hornet* mooring lines were connected to the pier at 0832 on July 26, after she had steamed 5,356 statute miles in support of the *Apollo 11* recovery operation. She returned with all her recovery personnel—*Plus Three*.

The MQF was towed onto the ship's starboard side elevator (#3). It was lifted by a giant moveable crane, its temporary wheels removed, and then lowered onto a USAF aircraft cargo loading vehicle. Captain Seiberlich joined Admiral McCain, Hawaii Governor John Burns, and Honolulu Mayor Frank Fasi for a brief pier-side welcoming ceremony.

Accompanied by a small motorcade, the MQF was driven over to Hickam AFB where flower leis were draped on the door handles. While waiting to be loaded into the C-141, Neil Armstrong saw a familiar face in the crowd and motioned him over to the MQF. He was USAF SSgt Eldridge Neal, the first pararescueman to reach astronauts Armstrong and Scott after the *Gemini 8* emergency landing in the Pacific Ocean in 1966.

The MQF was loaded into the hold of the giant cargo aircraft, which soon departed for Ellington Air Force Base. Following an eight-hour flight and brief reception in Houston, the astronauts were safely ensconced in NASA's Lunar Receiving Laboratory for the final two weeks of their quarantine period.

Fig 12K—This recent aerial view of Pearl Harbor shows Pier Bravo (X), the USS Arizona Memorial (Y) and Hickam AFB (Z). There is a ship moored to Pier Bravo in the exact spot where *Hornet* off-loaded the MQF in 1969.

Fig 12L—Two tugboats churn up mud as they nudge *Hornet* toward Pier Bravo and an expectant crowd of well-wishers. The CM *Columbia* can be seen on the starboard side of the flight deck near the bow (between the two rows of aircraft).

Fig 12M—A giant shipping crane lowered the MQF onto a 463L material handling vehicle, which itself is being hydraulically lowered to ground height. The crane's cables are being slackened to allow the workers to disconnect them while two TV cameras record the event for posterity.

Still on its dolly, the Command Module was craned off *Hornet* and towed to an aircraft hangar at Ford Island. For quarantine purposes, the hatch was never opened once the CM was disconnected from the plastic tunnel that connected it to the MQF aboard ship. It was emptied of unspent fuels by connecting external access ports to special ground support equipment. On July 29, the Command Module and the backup MQF were loaded into a USAF C-133, flown to Houston and then taken to the Johnson Space Center. The CM was placed in the Lunar Receiving Lab, where additional processing took place. It was released to the Smithsonian Institute once the quarantine period expired and has always been one of it most popular attractions.

Captain Seiberlich fully understood that *Hornet's* crew had spent over seven months on a WestPac cruise just prior to the eight weeks of the *Apollo 11* recovery mission. A few extra days had been built into the recovery schedule for contingencies, but the evolution had been completed sooner than expected. After off-loading all the NASA equipment and media personnel, *Hornet* could have spent a few more days in Hawaii. Seiberlich's leadership ability and command presence, however, were again in evidence. Prior to the splashdown, he had asked the ship's Executive Officer, Commander Lamb, to conduct a vote of the crew to see whether they wanted to spend more liberty time in Hawaii or immediately head back to Long Beach once the mission was complete. The vote was overwhelmingly for getting back to their families and friends. Twenty-four hours after tying up to Pier Bravo, and being the center of the world's attention, *Hornet* quietly slipped its mooring lines and headed east.

In Arlington Cemetery, the eternal flame over President's Kennedy's grave seemed to burn a little more brightly. The Navy had done its part.

Fig 12N—*Columbia* sits on its dolly in a hangar at Ford Island as a NASA technician prepares the spacecraft for its journey back to Houston.

Epilogue

Fig 13A—The insignia of *Apollo 12*.

After a brief respite in Long Beach, *Hornet* resumed its normal peacetime duties of performing carrier landing qualifications for naval aviators and researching Anti-Submarine Warfare tactics as part of the UPTIDE program.

Qualifying new naval aviators, or recertifying an experienced one, to do landings and launches from an aircraft carrier is know in the Navy as a "carqual." Making an arrested landing of a high-performance jet on the flight deck of a ship (known as a "trap") is a stressful process in the best of sun, wind and sea conditions. The pilot must approach the ship at the perfect angle and speed so the plane's tailhook can catch one of the arresting cables that cross the deck. A pilot can carefully adhere to all the procedures and rules, but a gust of wind or sudden ocean swell can cause his aircraft to bounce off the deck into the sea or be vaporized by striking the ramp at the fantail of the ship. Making a night landing can only be described as "magnificent terror," since the pilot can only see a few lights, not the flight deck. Even naval aviators who later became astronauts and walked on the Moon will admit their moment of highest mental clarity about life came during the first night trap.

On August 17, *Hornet* was selected as the Primary Recovery Ship for *Apollo 12*, the second lunar landing mission, scheduled to fly three months later. The astronaut crew for *Apollo 12*, Charles (Pete) Conrad (Mission Commander), Alan Bean (Lunar Module Pilot) and Richard Gordon (Command Module Pilot), were all naval aviators. Often referred to as the "All Navy Team," it was the only time in the Apollo program the entire crew came from the same branch of the service. *Hornet's* first major planning conference was held in early October just before an extensive, four-day INSURV (inspection and survey) of the ship by Rear Admiral J. D. Bulkeley. Detailed ship inspections such as these were highly disruptive yet the aging *Hornet* passed with flying colors.

Without the added complexity of a presidential visit to contemplate, the atmosphere was still highly professional but more relaxed. Other positive factors came into play as well. The mission profile called for a daylight splashdown giving the recovery forces, and TV media, long visual range. The extra planning time meant *Hornet* could install and integrate the SRN-9 TRANSIT satellite terminal into her navigation system. The lack of Moon germs found with the first lunar landing flight, and the discomfort created by the BIG suits, meant a slightly less ambitious quarantine protocol could be adopted. Almost the entire core recovery team was the same as it had been for *Apollo 11* with the primary difference being the UDT team. The "tribal knowledge" captured from the July recovery gave everyone a distinct advantage in terms of knowing what NASA's requirements and expectations were. Captain Seiberlich selected *Three More Like Before* as the motto for this recovery to again signify that safety was paramount.

Hornet steamed out of Long Beach on October 27 and pulled into Pearl Harbor four days later. After completing numerous SIMEXs locally, *Hornet* on-loaded the last of the recovery equipment and personnel. On November 10, she headed for her launch abort station below the equator, conducting SIMEXs along the way. *Hornet* was on-station for the launch and TLI abort events on November 16.

Early on November 24, *Hornet* arrived at the planned EOM location 375 miles east of Pago Pago (latitude 15.47° south, longitude 165.9° west). At 0958 local time, the Command Module *Yankee Clipper* splashed down only two-and-one-half miles from the ship, so the entire recovery operation was performed in full view of the media and ship's crew. The HS-4 Executive Officer, Commander Warren Aut, piloted the recovery helicopter while UDT-13 LTJG Ernie Jahncke handled the decontamination duties.

Once the astronauts were in the MQF, Admiral John McCain (CINC-PAC), Rear Admiral Donald Davis (Command, TF-130) and Captain Carl Seiberlich formally welcomed them back from their historic journey. On November 28, the MQF was off-loaded in Pearl Harbor and flown out of Hickam AFB so the "All Navy Team" could spend the rest of their quarantine period in Houston.

Hornet headed for Long Beach on November 29, arriving four days later. All hands enjoyed a well-deserved rest over the Christmas holidays. There was discussion about *Hornet* being selected as PRS for the *Apollo 13* mission, whose original launch date was set for March. However, *Hornet* was scheduled for decommissioning in June, which required months of deactivation and preservation

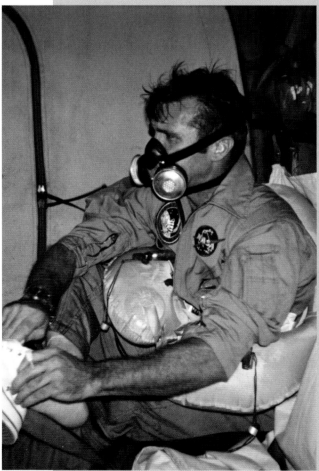

Fig 13B—*Apollo 12* Astronaut Dick Gordon is shown sitting in the cargo seat in back of Helicopter #66 after being hoisted up from the CM. Note the flight suit, respirator and tennis shoes that replaced the *Apollo 11* BIG suit for quarantine purposes.

Fig 13C—Commander of TF-130 RADM Donald Davis welcomes *Apollo 12* astronauts (l to r) Pete Conrad, Dick Gordon and Alan Bean back to Earth after their perfect mission.

work in preparation for long term storage. When NASA slipped the launch date to April, it was clear that *Hornet's* stellar participation in one of mankind's greatest achievements had come to an end.

In January of 1970, *Hornet* was involved in two ASW exercises designed to verify procedures for protecting a carrier battle group from enemy nuclear submarines. In February, during her last operational period at sea, *Hornet* conducted carrier qualifications for several air squadrons, recording 1,310 arrested landings on her flight deck. Finally, on February 20, she made her final arrested landing when an S-2E Tracker landed—*Hornet's* 115,445th career trap.

On March 30, *Hornet* departed Long Beach for her final trip to Bremerton, Washington. There she was placed in dry dock where her equipment, hatches, hull and other components were deactivated, removed, sealed and/or rust proofed for a lengthy mothball period. On June 26, 1970, *Hornet* was decommissioned from the U.S. Navy and placed in reserve status. Captain Carl Seiberlich was the last man to depart the ship, rendering a final hand salute to the old "Gray Ghost" of the Pacific.

For many years, *Hornet* remained quietly moored to a pier among several former warships, including the battleships USS *Missouri* (BB-63) and USS *New Jersey* (BB-62). Due to her contributions in war and peace, determined local community efforts succeeded in obtaining National Historic Landmark status for *Hornet* in December 1991.

Hornet remained in the Bremerton complex until late 1994 when it was towed to the San Francisco Bay Area to be scrapped at Hunters Point shipyard. While waiting to be dismantled by the "cutter's torch," she was temporarily moved to Alameda Naval Air Station in May 1995 to serve as the focal point for several months of celebrations commemorating the fiftieth anniversary of the end of World War II. Over 85,000 people visited her during the five-month period, generating a public ground-swell of support for saving the historic vessel.

The Aircraft Carrier Hornet Foundation was formed, raised funds and acquired the ship from the Navy as part of their museum ship donation program. In October 1998, the USS Hornet Museum was officially opened to the public at Alameda Point, in Alameda, California. The featured speaker at the grand opening ceremony was, appropriately enough, *Apollo 11* astronaut Buzz Aldrin. The museum maintains the world's only exhibit that specifically focuses on the splashdown and recovery of the early lunar landing Apollo missions when the Moon germ quarantine was in effect.

Fig 13D—The USS Hornet Museum is located at Pier 3 in Alameda Point just across the bay from San Francisco. It is moored next to the pier from which its predecessor, *Hornet* CV-8, sortied to launch the famous the Doolittle Raid against Japan in April 1942.

Appendix A
Personal Recollections

Neil Armstrong
Apollo 11 Mission Commander

Long before HS-4's Helicopter #66 ferried the *Apollo 11* astronauts back to *Hornet* in 1969, Neil Armstrong was intimately familiar with the flight deck of an Essex class aircraft carrier. In 1950, Armstrong became a naval aviator with Fighter Squadron 51 (VF-51), the Screaming Eagles. That summer they reported aboard the USS *Essex* (CV-9), the namesake for this famous World War II class of carriers. From August 1951 through March 1952, flying F9F-2 Panther jets, he logged seventy-eight combat missions in the Korean War.

In an email to this author, *Apollo 11* Mission Commander Armstrong remembers the recovery operation and ensuing quarantine:

The recovery, from my point of view, went extremely well. The Command Module had come to rest in the "Stable 2" (inverted) position, so we had the experience of going through the procedure to bring it to an upright position. Clancy [Hatleberg] and his UDT gang did an excellent job of securing the CM, getting us safely out of the spacecraft, completing

Fig 14A—NASA's official portrait of astronaut Neil Armstrong when he was selected Commander of the *Apollo 11* mission.

all the procedures for safeguarding Earth from lunar pathogens, and getting us into the basket to be lifted up to the recovery helicopter. The helicopter gave us an excellent ride and smooth landing on the *Hornet*. And the welcome by the *Hornet* and its crew, and the welcome by President Nixon, was memorable. We were convinced it had been a perfect recovery.

As a matter of fact, I didn't remember that the return to Pearl Harbor was two-and-one-half days. Dr. Carpentier took good care of us, the food and drink was excellent compared to our rations during the flight. We had a great deal of work to do getting our thoughts recorded as preparation for all the post-flight debriefings for which we were obligated.

The quarantine (in Houston) was a necessary nuisance but valuable for providing an atmosphere conducive to completing all our post-flight paperwork and interaction with later Apollo flight crews, systems specialists, flight controllers, etc. In view of the intense public interest in the flight, that would have been very difficult without the quarantine requirement. And the quarantine was a constant reminder that *Apollo 11* had been a success in reaching the national goal of landing men on the surface of the Moon and returning them safely to Earth.

Fig 14B—This rare photo shows Neil Armstrong during the ticker tape parade given to the *Apollo 11* astronauts in August 1969 just after their release from quarantine.

John Stonesifer
NASA Quarantine Manager for *Apollo 11*

During the Apollo Program, personnel of the NASA-JSC Landing and Recovery Division and other NASA personnel (medical, photographic, public affairs, etc) accompanied the various DoD forces participating in each Apollo recovery mission. The Team Leader embarked on the primary recovery ship had several responsibilities including:
- Coordinating all NASA-oriented activities in the primary landing area.
- Managing the various NASA team leaders.
- Acting as the point of contact between the NASA recovery team and the ship's command.
- Coordinating special briefings for all recovery forces and the media.

John Stonesifer was the Team Leader for twenty-one spacecraft recovery missions. These included most of the Mercury series, most of the Gemini series, one of the Apollo unmanned flights, and *Apollo 8*, *11* and *12* manned missions. He was in the Mission Control Center for the others. He also went to sea to check out the special medical vans used put on the recovery ship for the Skylab missions.

John recalls:

There were always unique happenings and challenges aboard the different ships. The biggest challenge was introducing the newly assigned DoD teams to recovery operations and procedures as new missions were scheduled. Outstanding memories were many: the inevitable delays in space launches permitting more time at sea for recovery simulations, developing lifetime

Fig 14C—John Stonesifer chats with CDR Warren Aut of HS-4, the primary recovery pilot for the *Apollo 12* mission, beside the fuselage of the famed Helicopter #66

acquaintances with devoted DoD personnel, working with the wonderful and dedicated NASA personnel as part of my team, providing solutions to the occasional difficulties of working with the press corps who pushed for priorities over fellow press members and applying new technologies as programs progressed over the years.

I was the Quarantine Manager for the *Apollo 11* mission. It is my most memorable recovery because of its historic nature and the challenges I faced making sure the isolation/quarantine was completely successful. The ICBC was concerned about the possibility that unknown pathogens returned from the Moon could result in catastrophic consequences here on Earth if the isolation/quarantine process failed.

I enjoyed briefing President Nixon and Admiral McCain about the importance of the quarantine program and the precautions we must take. At one point in the conversation, I explained that two NASA folks had volunteered to enter quarantine with the astronauts, one a doctor to examine the astronauts and administer care if required, and an MQF engineer to keep all the systems operating until they got back to Houston. President Nixon put his hands on his hips and said, "When I was in the Navy I never volunteered for anything." We all had a great chuckle!

I was very pleased the recovery operation on *Hornet* went very smoothly. But we still had much to go through after we left *Hornet*. I sweated every moment till we got the MQF mated to the Lunar Receiving Lab in Houston. The transfer from the ship to the aircraft at Hickam AFB went smoothly and here again our SIMEXs, (yes, we even had rehearsed the loading and connections in previous weeks) proved beneficial for all and worked as planned. It was very rewarding aboard our C-141 to receive messages from other aircraft and ships along our flight route. They apparently were aware of what was occurring and radioed "well done," "Godspeed" and "congratulations" to the astronauts.

The arrival in Houston was exciting. We had a police escort from Ellington Field to JSC, a special greeting event with the astronaut's wives at the MQF window and the mating of the MQF to the LRL so the astronauts could enter their final quarantine "home." Once that was completed and the MQF was buttoned up, my assignment for that mission was complete (except for the ubiquitous post mission reports).

Fig 14D—John Stonesifer gives President Nixon a quick overview of the quarantine program for *Apollo 11* about one-half hour before the recovery operation is scheduled to begin.

I took great comfort that everything we had rehearsed over the previous months was carried out perfectly by the entire team. I often felt what a failure it would have been if the last parts of the mission, the quarantine—that part least understood by the public—had been compromised.

We reviewed every aspect of the *Apollo 11* operation and concluded we're ready to do it again for *Apollo 12,* only four months from then. We knew we had done what the ICBC had directed.

Another recovery that really stands out in my mind is *Apollo 8*. The recovery took place over the Christmas holidays—way down in the South Pacific. There was *elation* as it involved a spacecraft that carried humans on the first trip around the Moon, yet there was great *tension* waiting for that single rocket to fire on the back side of the Moon to bring it home. Also while the USS *Yorktown* was en route to the recovery area, Twentieth Century Fox was filming a segment of the movie *Tora, Tora, Tora* onboard. Thirty propeller-driven aircraft, mostly former U.S. Navy training planes painted to resemble Japanese warplanes, were launched off the flight deck. Shortly after we left San Diego for Pearl Harbor, these planes took off to simulate the attack on Hawaii in 1941 that brought the U.S. into World War II.

Another one of special note is MA-7 [*Aurora 7*] when Scott Carpenter's Mercury capsule landed 250 miles beyond his planned landing location. Here the results of going through a "what if" scenario paid off—something I am proud that we, NASA and the deployed forces, continually discussed and planned. The new SH-3A Sea King helicopters aboard the PRS were not considered part of the recovery force. But as a result of discussing contingency scenarios for certain conditions, they were quickly deployed and brought about a rapid success to an unusual recovery operation.

Captain Charles B. Smiley
HS-4 Recovery Pilot for *Apollo 10* and *Apollo 13*

Charles "Chuck" Smiley became a naval aviator in September 1953 and initially served with Patrol Squadron 19 and qualified to fly helicopters in 1960. The Navy assigned him as an exchange pilot with the Royal Australian Navy in 1962 and 1963.

In the fall of 1968, Commander Smiley became the Executive Officer of Anti-Submarine (ASW) Helicopter Squadron 4 (HS-4) based at Imperial Beach, California, an auxiliary airfield just outside San Diego. HS-4 was comprised of sixteen SH-3D Sea King helicopters with fifty officers and about 250 enlisted personnel. Commander Smiley was the primary recovery pilot for both the *Apollo 10* and *Apollo 13* missions.

He remembers his initial introduction to the Apollo recovery program:

Commander Don Jones, the Commanding Officer of HS-4, walked into my office in November 1968 and simply said, "We've been selected as the primary recovery squadron for the *Apollo 8* mission."

We knew that NASA and the Navy had been using Sea Kings during the final preparations (or "work-up missions") in the Gulf of Mexico, but this was "big time." *Apollo 8* would be the first time humans left the field of Earth's gravity and also the first time humans had ever seen the far side of the Moon. The lunar orbit meant it would result in a Pacific Ocean splashdown and HS-4 was going to play a crucial role in the "Main Event"!

As the plans began to unfold, it was determined that Don would deploy aboard USS *Yorktown* with half the squadron personnel and eight aircraft. I would remain behind and continue the ASW work-up, training for our next combat deployment.

Fig 14F—CDR Chuck Smiley in flight gear during one of the Apollo recovery missions.

While we had never split the squadron before, we needed to maintain our anti-submarine operational commitments while executing a flawless Apollo recovery. We were too busy to worry about it—we just found a way to get everything accomplished.

Don, of course, performed a textbook recovery operation. Those of us still at the base watched it on TV. We were wondering what must be going through his mind? Thoughts like, "Just do your job . . . no slip-ups. Just like any other recovery . . . except . . . a worldwide TV audience, so its gotta be perfect!" And he was.

The *Apollo 9* flight in March 1969 was an Earth orbit mission, which meant an Atlantic Ocean splashdown. George Rankin and HS-3 operating off the USS *Guadalcanal* executed a great recovery.

Then, in April 1969, the word came down that HS-4 had been selected to handle the *Apollo 10* astronaut recovery. Certainly, the easy way to do that would have been to use the same team that recovered *Apollo 8* just six months before. But that was not Don's style. As CO, he wanted to ensure that everyone had an opportunity to accomplish a once-in-a-lifetime mission; to be involved in a very special experience. He always had a larger perspective on life, which may be a key reason he made Vice Admiral and became the Commander, 3rd Fleet, later on in his naval career.

As a result, I was selected to embark the HS-4 detachment assigned to the *Apollo 10* recovery mission and to pilot the astronaut recovery helicopter. Just before dawn on May 26, 1969, I lifted Helicopter #66, call-sign *Recovery Three*, off the deck of the USS *Princeton*. We maintained a 5,000-foot high orbit around the ship, counting down the seconds until the spacecraft's re-entry. Suddenly, a speck of light appeared in the sky just exactly where we had been told it would be by the NASA folks. The speck grew a tail of flame and appeared to be climbing, which was an optical illusion. It was actually descending at very high speed through the upper atmosphere. The object continued until it was directly overhead. Then the flame flickered out, which meant it was in a vertical or straight down trajectory!

A thought momentarily flashed through my mind that the spacecraft might actually be directly overhead and plummet right on top of us. Ah, a billion to one chance? No time for that. Relax and get it done. Shortly afterward, I got my first view of the spacecraft's Earth Landing System with the flashing light, three giant

Fig 14G—CDR Smiley points out the spacecraft decals on the fuselage of Helicopter #66 to *Hornet* CO Captain Carl Seiberlich.

parachutes and the Command Module silhouetted against the pre-dawn sky. From the top of the chutes to the bottom of the spacecraft is roughly 210 feet, equivalent to a twenty-story building descending through the sky just a mile away. What a thrilling sight that was.

SPLASHDOWN!

The *Apollo 10* Command Module, with the radio call-sign *Charlie Brown* (named after the *Peanuts* cartoon character), hit the water only three or so miles from the Primary Recovery Ship. It remained in a Stable 1 position with its apex up. Our recovery plan called for immediately ensuring the space crew was in good shape and then waiting a few minutes for daylight before we proceeded with the actual recovery. I held *Recovery Three* in a hover about forty feet above the water and had a front-row view. Very shortly after splashdown, we saw light emanating from the spacecraft hatch and figures came into view. I turned to my copilot Scott Walker and said "Do you realize those three guys just came back from the Moon?"

We deployed the UDT team, who removed the main parachutes from the recovery area. Having one of those huge orange and white chutes "blossom" into the helicopter tail rotor would ruin a pilot's whole day. The UDT swimmers then installed the flotation collar and attached a life raft to the front of the Command Module, right on schedule. LTJG Wes Chesser opened the hatch and assisted the astronauts into the life raft.

It was our show then.

Chief Petty Officer Glen Slider was positioned at the starboard cargo hatch in the rear of our Sea King. His role was to operate the hoist and talk me into the correct hover position. As pilot, I had to focus at a point in the water about fifty yards ahead to detect altitude and relative motion in order to maintain a stable hover position. Scotty took care of everything inside the cockpit. All I had to do was follow directions, just as we'd practiced in many hours of simulated exercises . . . except this time was for real. Stafford, Cernan and Young, not stand-ins, were climbing into the Billy Pugh net one-by-one, and they've just returned from the Moon!

Within a few minutes, all three were safely aboard. We make a circuitous flight back to the USS *Princeton*, so the space crew could change out of their Constant Wear Garments and slip into their light-blue NASA flight suits. With the eyes of the world on America's space program, they needed to look good for TV.

Once we received the call from the ship "Cleared to land on Spot 5," Scotty and I went over the landing checklist and then settled Helo #66 gently down on the flight deck. We shut down the engines and put on the rotor brake while taking note of the mass of humanity around us as well as the Navy brass hanging out along "Vultures Row" on the island. The *Apollo 10* crew disembarked and were welcomed by a traditional Navy ceremony on the flight deck.

I turn to Scotty for our usual macabre sign off. . . "Well, we cheated death again."

He just smiled.

Clancy Hatleberg
Officer in Charge — *Apollo 11* UDT Teams

Clancy Hatleberg attended Dartmouth College on an NROTC scholarship. He graduated in June 1965 with a B.A. degree and was commissioned an ensign in the U.S. Navy. His goal was to join the Underwater Demolition Teams, but the Navy required that he serve aboard a ship before he was eligible for UDT training. He received his orders to the UDT school in June 1966.

In mid-May 1969, while UDT-11 was training to recover *Apollo 10*, Lieutenant Hatleberg was selected to be the decontamination swimmer for *Apollo 11*.

Hatleberg recalls what transpired after that:

I was pulled off the *Apollo 10* recovery a few days before its splashdown and traveled to Houston to train with the astronauts. We trained in Galveston Bay for a morning, after a day of work-up at Johnson Space Center.

The work-up focused mainly on the new decontamination procedures, washing down the astronauts, then the CM, collar, and finally the raft. I only practiced with betadine, a surgical scrub, for the whole procedure until doing SIMEXs aboard *Hornet* on the way to the recovery area for *Apollo 11*. Then we learned that betadine is a wetting solution and made the Biological Isolation Garments (BIG suits) permeable. Plus, betadine made the rubber life rafts slippery and difficult for me to work in. The procedure was then changed to include bleach (sodium hypochlorite) for washing down the astronaut's BIG suits. I enjoyed the opportunity to interact with the astronauts and the NASA recovery team in Houston.

The UDT-11 primary recovery team was led by LTJG Wes Chesser and included QM3 Michael Mallory and SN John Wolfram. Members of this team had been instrumental in the successful recovery of *Apollo 10*. UDT-12 provided the secondary recovery team, which was also airborne during the actual recovery.

Fig 14H—Watching intently from a helicopter hatch, Clancy Hatleberg (in white helmet) observes one of the many training exercises (SIMEXs) conducted by the UDT teams to ensure a flawless recovery of *Apollo 11*.

The third recovery team, also from UDT-11, was the standby team that waited on the flight deck in case of an emergency. Just to confuse matters, the primary team was loaded onto the *Swim 2* helicopter while the secondary team boarded *Swim 1*.

NASA provided a practice Command Module, usually referred to as a boilerplate, for training use by UDT personnel. The teams continued extensive preparation while at Coronado, becoming familiar with recovery procedures, spacecraft components, special equipment, etc. They also performed mock-up recovery exercises in San Diego harbor.

The UDT recovery group embarked on *Hornet* as soon as it arrived in Pearl Harbor (July 3), stowing its equipment in Hangar Bay #2. During the next few weeks of SIMEX training, the UDT teams practiced the recovery procedures twelve times, with each team getting at least one day and one night SIMEX. During this time-frame, the scrubbing agent for the astronaut BIG suits was changed to sodium hypochlorite, a better solution for ensuring the quarantine effectiveness and a safer raft environment.

On splashdown day, *Columbia* landed thirteen miles from *Hornet*, while it was still semi-dark, in a higher-than-expected sea-state and in a Stable 2 position (apex down). After it had righted itself, Chesser, Mallory and Wolfram were able to efficiently complete their CM stabilization tasks. *Recovery One* then maneuvered into position so Hatleberg, the "decon" swimmer, could perform his part of the recovery process.

Hatleberg recalls the historic events that followed:

My main concern was with the high sea-state—these were the worst conditions we had worked in. After the astronauts donned BIG suits, there would have been a major problem maintaining the quarantine had one of them had gotten sick.

I wasn't so much worried about following the decontamination procedures that NASA had drilled us on—that process had become second nature. But, on the night before the recovery, I had been asked to do three things that I hadn't practiced:

- Check the CM vents on the top to make sure they were closed and not still venting into the Earth's atmosphere

Fig 14I—During the NASA water egress training for the *Apollo 11* astronauts in the Gulf of Mexico, Hatleberg also practiced decontaminating the Command Module with betadine.

- Make sure the tape on the front of the astronauts BIG suit filters was removed so they could breath properly
- Ensure the water wings on each BIG were inflated in case an astronaut fell out of the raft.

I couldn't bring a check-list with me, so while I was waiting in the helicopter for the CM to splash down, I used a grease pencil to write "vent, tape, inflate" on the front of my face mask. Later, when I put it on to jump out, I realized it read backwards from the inside. But, at least I didn't forget to do them.

As soon as *Recovery One* was close to *Columbia*, I jumped into the water and swam to the upwind raft. The helo crew then lowered the BIG suits, decontamination canisters and wiping mitt to me via Billy Pugh net. I put on my BIG suit, which filtered the air I breathed in, and was pulled over to the decontamination raft attached to the CM. There I secured the decontamination canisters, one with bleach to scrub the astronauts and the other with betadine to wash down the CM, collar, and raft.

When all my equipment was ready, I gave a thumbs-up signal to the helo, requesting them to have the astronauts open the CM hatch. I handed the astronauts a bag with the three remaining BIG suits, which filtered the air being breathed out, and the hatch was closed. After the astronauts had donned their BIG suits, the hatch was opened again and they exited the CM into the decon raft one at a time.

After they were all safely in the raft, with water wings inflated, the first problem happened. The main CM hatch was left in the "open" cycle and I couldn't get it to latch shut. I tried a few times and then looked into the raft with my arms up in the "I need help" position. Mike Collins had already realized there was a problem. He came over and recycled the hatch so I could close it properly.

Then, I started washing the astronauts down with the mitt, using the bleach. This went well, even in the high sea-state, until the wave-of-the-day broke over the raft. I was down low washing one of the astronaut's booties when the wave collapsed over us and knocked me flat. Before I got up, I figured I'd be the only one left in the raft and have to swim over the horizon to Australia as per Dr. Stullken's prior instructions when he told me failure was not an option.

Anyway, we all stayed in the raft and after the decontamination was completed, each astronaut was hoisted up to the recovery helo in the Billy Pugh net. Once they were aboard the helo, I decontaminated the CM, collar, and recovery raft. Then I took off my BIG suit, secured it in the recovery raft along with all the decontamination equipment. After waiting thirty minutes to be certain the decontaminate solution had been effective, I slashed the raft to sink it all to the bottom of the Pacific.

Hatleberg remained with the three other UDT swimmers to prepare the CM for its hoisting aboard the USS *Hornet*, which occurred almost exactly three hours after it had hit the water.

Don Blair
Mutual Broadcasting System Broadcaster

Don Blair was born and raised in New Jersey. After a two year hitch in the U.S. Army, he began a nearly fifty year career in radio and television. From 1965 to 1989 he worked for all four radio networks in New York City, writing and delivering an estimated 25,000 newscasts during that period. Blair was the live radio reporter for five spacecraft recoveries and is the author of the 2004 book called *Splashdown: NASA and the Navy*, a delightful remembrance of these experiences.

Blair recalls:

During the 1960's, there were but four networks in existence and one of them, mine, is one that many people don't remember—the Mutual Broadcasting System (radio only).

As NASA became a household word, each new mission was reported to the public by NBC, CBS or ABC on television, with my Mutual network pitching in with radio coverage. When Alan Shepard headed into sub-orbital space in 1961 in his tiny Mercury space capsule, the recovery reporting was handled by just two radio/TV correspondents and a small number of representatives of the print media. We were considered a "pool," meaning that a few of us reported for everybody else.

With each passing day at sea aboard the Prime Recovery Ship, a correspondent from NBC, for instance, would also write and deliver a daily radio report for CBS, while the guy from ABC would file a similar report for Mutual. Naturally he would change his sign-off for the other network he was serving. These reports followed days filled with briefings

Fig 14J—Don Blair reporting to his worldwide radio audience from a lookout station high on the aircraft carrier superstructure.

by NASA and Navy personnel regarding the constant recovery rehearsals using boilerplate or "dummy" spacecraft and the on-board UDT team members or—as we called them—frogmen. Our daily stories were also augmented with ongoing briefings on sea conditions in the planned splashdown area with more than occasional changes due to unacceptable weather conditions cropping up.

So it was two correspondents working for the four networks but . . . on the actual recovery day, that rule went out the window. Both men would share the live splashdown coverage and their descriptions would go out to all the networks as one. Back home you would read, on your TV screen, "Don Blair reporting from the USS *Wasp*" or Bill Ryan or Dallas Townsend—to name just a few who were given those great assignments over the years. Those assignments, by the way, went to networks picked from a hat or an ashtray at news directors meetings in Manhattan some weeks before every mission. That was true not just for the reporters but for the technical crews—which network would send the TV cameras, the directors, lighting and sound personnel, etc. In my case, I just happened to draw three straight assignments, *Gemini 9*, *Gemini 10* and *Gemini 11* in 1966—purely through the luck of the draw.

The one mission all of us wanted was, of course, *Apollo 11*—first landing on the Moon. As 1969 began, the news directors wisely chose to send a correspondent from each of the TV networks—Ron Nessen of NBC, Dallas Townsend from CBS and Keith McBee of ABC—in addition to myself. This was an acknowledgment of the biggest story of our time. We all went about our business as on earlier missions, digging up stories for daily radio prior to recovery day, interviews, news briefings, human interest stories. However, on July 24th, 1969 with astronauts Neil Armstrong, Buzz Aldrin and Michael Collins racing home from the Moon, the broadcast media fanned out over the decks of the USS *Hornet*, each stationing themselves in areas complimenting the other's view of space history.

For myself, I climbed the stairs in *Hornet's* island to the 06 level, where I had camped out on all my previous mission assignments. This is the highest level most personnel can go to. It's near the Navigation Bridge and the Combat Information Center where the top officials from both NASA and the Navy gathered to process the streams of information coming in from Houston and the returning *Apollo 11* spacecraft. I had selected a small half-moon cubicle from which a sailor using semaphore flags would normally be stationed to signal other ships in a fleet during naval exercises. It was a superb position and I used it on all five of my downrange assignments.

On that historic day, LCDR Smith was my go-between. He consistently handed me little notes of the vital information being as-

Fig 14K—A ship's public affairs officer assigned to each TV and radio broadcast reporter would listen to the cryptic Navy voice traffic and explain what activity was taking place.

sembled in the CIC, which allowed me to keep my radio audience, world-wide, abreast of the latest developments. With clearly defined breaks in our coverage—perhaps ten or fifteen minutes at a time when information was not that vital or current—I stayed on the air for nearly five hours solo. The three-ring notebook under my arm was rarely used. Picture a person doing sports play-by-play. If he is doing his job and reporting what he sees before him, or a "spotter" is conveying to him, there is little time to go looking through notes. You are an eyewitness to history and need to relay the unfolding drama to your audience as it happens.

When *Apollo 11* splashed down that morning some thirteen miles away, my TV colleagues thought the CO of *Hornet* had misjudged the distance. As Captain Carl Seiberlich was to explain many times in the passing years, he had the President of the United States, the Secretary of State, the head of NASA and many valuable lives onboard his ship. There was an incoming spacecraft that might be spewing toxic chemicals or "Moon germs" into the air as it floated down nearby. Captain Seiberlich erred on the side of safety and was right in doing so.

The moment the *Apollo 11* astronauts were safely ensconced in the MQF, my broadcast assignment was over. My producer said we probably had the largest radio audience of all time. I was very happy to hear that and took my cameras over to the catwalks surrounding *Hornet's* flight deck, to witness and photograph the retrieval of the Command Module. I was now a happy tourist, with some of the top photographers and newsmen on Earth all around me. A photo of *Apollo 11*, which I still use in my lectures, was almost identical to the one used by Look magazine in a special Lunar edition that appeared about a week later.

Members of the media were quartered above the hangar deck level, just below the flight deck, in an area called "Officer's Country." Each room had two comfortable steel-framed bunks, a desk, two chairs, a sink, medicine cabinet and large lockers. The shower/bathroom was just a few steps away. The food was uniformly good to excellent on all my splashdown assignments. On the night the *Apollo 11* astronauts were safely back on Earth, the celebratory dinner menu would have done any fine hotel proud. I still have that menu, but more importantly, I will always have the memories of one of the greatest assignments any newsman could ever hope for. The Navy and the USS *Hornet*—none better.

Milt Putnam
HS-4 Photographer

"I joined the Navy with an intention of pursuing photojournalism as my specialty. It took several years for me to finally move into a Photographers Mate billet. Being a photographer and a qualified aircrewman with an air group is best with helicopter squadrons. With helicopters, the photographer is always where the action is because the Navy needs pictures of everything.

"I was aboard the USS *Yorktown* off Vietnam in 1968 when I was picked for the *Apollo 8* photo assignment with HS-4. I had to do an at-sea highline transfer from the USS *Hassayampa* over to *Yorktown* to get into the TF-130 operation for that recovery.

"There were many photographers onboard the USS *Hornet* during the *Apollo 11* mission. The ship had at least 12-15 of its own and many of the ship's crew had cameras as well. *Life Magazine*, *National Geographic*, the Associated Press, United Press International and dozens of other news outlets also had photographers there.

"I couldn't sleep the night before and the HS-4 flight crews arose at about 3 a.m. NASA photographer Lee Jones and I were the only people shooting from the *Photo* helicopter at the splashdown scene and he shot mostly 16mm motion picture film. We were both strapped into safety harnesses and sat in the open cargo hatchway getting the best view of the recovery operation possible. One of my thoughts at this time was how lucky I was to be shooting the biggest photo assignment man has ever recorded on film. I was scared half to death, but the fear went away after the first shutter click. As the only still camera shooter flying the helicopter portion of the recovery mission, I knew I had to get pictures of everything. I wanted to prove to all those magazine and news photographers back on *Hornet* that the Navy had picked the right person for this job.

Fig 14L—Milt Putnam is shown standing next to Helicopter #66 in his flight gear preparing for a simulated recovery exercise.

"My primary photographic mission was to shoot the recovery both from the air and also in the hangar bay area. After the astronaut retrieval was completed, the *Photo* helo sped back to the *Hornet* and landed first so I would be back aboard before *Recovery One* landed. I was loaded down with several cameras strapped around my neck and body, plus fifty or so exposed rolls of film in my flight suit pockets along with about ten to fifteen rolls of unexposed film. President Nixon and everybody except God was standing up on the island and the world was watching on TV. Just as we touched down on the flight deck, I jumped out of the helicopter and, unfortunately, was still strapped into my safety belt! Somehow I didn't fall over, but I was hanging there with my feet only an inch or so above the deck for what seemed an eternity. I unsnapped the belt as quickly as I could and no one knew what had happened (I hope).

"Most of the film I shot that day was in black and white and was developed aboard *Hornet* by the Associated Press guys. They selected what they wanted to place on the news wire and filed those images at AP. Motion picture film footage shot by two Navy Combat Cameramen assigned to *Hornet* was shipped to the Naval Photo Center at DC for developing.

"Being a Navy photographer opened many doors for me both in the military service and later in my civilian career. After photographing the first three Apollo lunar recoveries, *Apollo 8*, *10*, and *11*, I had a reputation as one of the Navy's best shooters. Public Affairs Officers throughout the country would call my command, requesting that I shoot big events for them. These assignments included the Queen of England's visit to Boston in 1976, a three-month expedition to Antarctica with the National Science Foundation in 1972-1973, the return of the POW's from Vietnam, and earthquake relief in Peru.

"My life changed after *Apollo 11* because the pressure was always on to do my best and come up with pictures that no other Navy photographers were getting. I was always searching for new angles to make my photos different from everyone else. A few years after Apollo, the Navy designated me a photojournalist without my attending the required course at Syracuse University.

"I had no problem finding a job after retirement from the Navy. The Apollo recoveries boosted

Fig 14M—A rare picture showing the photographer Milt Putnam sharing some words with President Nixon as the latter was preparing to board his helicopter and leave *Hornet*.

my military career and, almost forty years later, people still comment on my being the guy who photographed Neil Armstrong's return from the Moon. Pictures I took show up everywhere in history books, books about the U.S. Navy and the space program.

"I can still feel what it was like sitting in the hatch of the Sea King helicopter, loaded down with cameras, capturing history on film. The shaking of the helicopter and the rotor blades pushing the balmy air through the hatch with history unfolding in front of my camera lenses is something I'll never forget."

Rear Admiral Carl J. Seiberlich
Commanding Officer USS *Hornet* (CVS-12)
Commander of *Apollo 11* Primary Recovery Forces

Carl J. Seiberlich graduated in 1943 from the newly established U.S. Merchant Marine Academy at King's Point, New York with a Bachelor of Science in Marine Transportation.

During World War II, he served as navigator aboard the destroyer USS *Mayo* (DD-422) in both the Atlantic and Pacific theaters. The *Mayo* was anchored in Tokyo Bay, close to the USS *Missouri*, when the formal Japanese surrender was signed on September 2, 1945.

After the war, he became involved in naval aviation, first flying lighter-than-air craft (i.e., blimps). In 1949, he worked on Anti-Submarine Warfare (ASW) ship-to-airship tactics, perfecting fleet techniques that later became standard practice for the Navy. For this effort, he was awarded the 1951 Harmon International Trophy by President Truman.

His first deep-draft ship command was the USS *Salamonie* (AO-26), which he decommissioned in late 1968. In May 1969, Captain Seiberlich took command of the USS *Hornet* (CVS-12) in Long Beach, California, upon its return from combat duty off the coast of Vietnam. Almost immediately, *Hornet* was selected as the Primary Recovery Ship for the first lunar landing mission, *Apollo 11*.

Captain Seiberlich became Commander of the Primary Landing Area Recovery Forces in the Pacific Ocean, and was designated the "on scene" boss for the entire recovery operation. This action was significantly more complicated than any previous one because of the Moon germ issue, the presence of President Nixon and vast media attention.

Fig 14N—This official Navy portrait was taken in June 1969, immediately after Captain Seiberlich assumed command of the USS *Hornet*.

Apollo 11 splashed down in the middle of the Pacific Ocean on July 24, 1969. Seiberlich and his crew executed a flawless recovery of astronauts, spacecraft and precious Moon rocks.

Exactly four months later, *Hornet* repeated this stellar performance as the primary recovery ship for the second lunar landing mission, *Apollo 12*. On November 24, 1969 the spacecraft *Yankee Clipper*, with its all-Navy team of astronauts, splashed down only two miles from *Hornet*.

In the process of converting its fleet to nuclear power, the Navy decided the WWII aircraft carrier's service time had run out. In June 1970, Captain Seiberlich presided over the decommissioning ceremony of *Hornet* in Bremerton, Washington.

After selection for flag rank, Admiral Seiberlich was assigned as Commander, Anti-Submarine Warfare Group Three with the USS *Ticonderoga* (CVS-14) as his flagship. He was a key participant in an ASW exercise called Project UPTIDE, which helped modernize Navy doctrine for defending a carrier battle group against nuclear submarines. It is of note that Rear Admiral Seiberlich was the first USMMA graduate to achieve flag rank in the U.S. Navy.

RADM Carl Seiberlich passed away in March 2006 and was buried at Arlington National Cemetery with full military honors.

RADM Carl J. Seiberlich, USN (Ret.) was interviewed in December 2003 by Hornet Museum personnel. Even at age eighty-one, he recalled vividly details about these historic events.

Fig 140—Captain Seiberlich escorts President Nixon, Admiral McCain and other dignitaries across *Hornet's* flight deck after the recovery of the *Apollo 11* astronauts.

HM: Why was *Hornet* selected for the recovery of *Apollo 11*?

CJS: NASA wanted a "four-screw" aircraft carrier to manage the recovery operation. Our fully redundant propulsion and power systems provided insurance in case of a failure. Also, *Hornet*'s duty schedule fit the timing of the mission launch and recovery windows. A highly maneuverable ship, *Hornet* would be able to take various contingency actions if there was a Moon germ issue or personnel contamination. Moon germs were a big concern at the time, and we had to protect citizens against this potential danger.

HM: What were some of the early issues you tackled?

CJS: During the planning stage, we felt the ship should always be upwind from the CM *Columbia* when it re-entered the Earth's atmosphere. This ensured that when its internal environment was purged during the descent (to equalize pressure inside), we would not endanger the ship or crew from a possible release of Moon germs or toxic spacecraft chemicals into the air.

    We held our initial practices in Hawaii to determine how the capsule would float, both by itself and once the flotation collar was attached. We also studied how to get the isolation garments into the capsule for the astronauts to don, and to understand the process of safely hoisting a bobbing five-ton capsule aboard a 44,000-ton ship in the middle of the Pacific Ocean. After these practices in Hawaii, we boarded the NASA personnel and lots of media folks and headed to the designated splashdown area. On the way, we held additional simulated recovery exercises (SIMEXs) to stay sharp and provide backup personnel for every contingency. The final success is a reflection of having great team leadership from many people such as Don Stullken (NASA), John Stonesifer (NASA), Don Jones (HS-4) and Clancy Hatleberg (UDT-11).

HM: What were some of the issues you encountered along the way?

CJS: Actually, we had mostly small issues right before the operation. We needed to have a ball cap with our slogan "Hornet Plus Three" for the senior American executive to welcome home the astronauts. Originally, we were told Vice President Agnew was coming, so we had a cap made for him. But, as we were about to leave San Diego for Hawaii, I learned that President Nixon was going to attend. We didn't even know his cap size. So we called a cap maker in New York who sent several different sized Presidential "Hornet Plus Three" caps, which were waiting for us at the pier in Pearl Harbor when we arrived!

    We had another issue when a couple hundred newspaper, TV and radio personnel reported onboard to cover the event. I briefed them about conserving water, since the ship's boilers had first preference on the desalinated water supply to keep the ship moving. But, within 48 hours of their arrival, we had run out of water. We explained to them how to take a "Navy shower," and after that, we didn't have a problem. In an effort to get the best photo, some of the media got into fights with each other so we made chalk outlines on the deck to show each one where to stand during the recovery operation.

HM: As the day of the *Apollo 11* splashdown approached, what actions did *Hornet* take?

CJS: The splashdown was scheduled for 5:40 a.m., well before sun-up, so there was no horizon or satellite navigation system by which to navigate or judge our position. After arriving at the original station, we examined the ocean floor with the fathometer for mountains to serve as static guideposts, along with taking celestial readings. We found four prominent seamounts and I named them *Armstrong*, *Aldrin*, *Collins* and *Apollo 11*. I gained a reputation for doing "ocean exploration" from a news reporter, but it was just a system devised to maintain our proper location in the splashdown area with very few navigational aids.

On the day before the recovery, there were stormy conditions in the primary target area. I had been cautioned by NASA that if two of the spacecraft's parachutes collapsed (due to high wind or other factors), the astronauts would not survive the landing. I advised Houston to relocate the recovery area away from the storms. Upon approval of the new recovery site, *Hornet* steamed at near flank speed (25 knots) and used dead reckoning navigation to get to the new spot in time, which was over two hundred miles northeast. While NASA doctrine allotted five miles of error on either side of the splashdown point, I used that entire margin of error to remain upwind eleven miles from the impact point. I couldn't risk getting the President contaminated with Moon germs, and it was more important for the helicopters to be at the splashdown point than the ship!

Some of the media was critical of this, especially TV, but they also didn't completely grasp the gravity of the Moon germ concern. It was a very complicated task to coordinate all the variables involved, and trade-offs were made in favor of safety.

HM: What was the most memorable experience you had during the recovery?

CJS: President Nixon was on the flight deck preparing to leave for his worldwide tour and wanted to thank the crew for doing an excellent job. He ran over to a "purple shirt" deckhand and told the crewman, "Great job today." This enlisted man simply replied, "Sir, we're *Hornet*." To me, that summed it up! The President knew the tradition of high standards associated with *Hornet*'s name since the first U.S. vessel of that name, commissioned in 1775. The President turned to me and said "Tell the crew that today, they have met *Hornet's* standard." Our success is a direct reflection of having a crew of dedicated men doing their job to the best of their ability.

One of the most important, and virtually unheralded, members of our crew was the young man who ran the B & A crane and hoisted the capsule onto the ship. For a few minutes, he was basically in charge of the $25 billion *Apollo 11* mission, and had to make sure the Moon rocks didn't sink to the bottom of the ocean. There was a lot of pressure to do things right, but to the *Hornet* crew, it was just another day of doing their job well. We didn't spend time thinking about how *Hornet* was going to be part of world history until it was all over.

HM: What do you consider *Hornet*'s legacy today?

CJS: The heritage of the name *Hornet* extends back to the Revolutionary Navy. There have been eight ships named *Hornet* as well as the current naval strike fighter (F/A-18). One can view much of the history of the U.S. through this lens. *Hornet* CV-8 altered the course of history by its actions regarding the Doolittle Raid and the Battle of Midway in World War II.

Hornet CV-12 not only has excellent World War II and Vietnam War combat records, but is arguably the most world-famous U.S. aircraft carrier because of the role she played in the *Apollo 11* and *Apollo 12* recoveries. Remember, President Kennedy said the job wasn't done until those lunar explorers were returned safely to Earth. This key role in the space program needs to be passed on to future generations, and everything else she participated in provides an excellent context for that lesson. As a museum and tourist attraction, the USS Hornet Museum is now a vehicle to reach out and educate many generations to come.

Final Thoughts

I promised RADM Seiberlich that when the book was finished, he could have the last word. We met many times and each interaction followed the same general pattern. He would patiently listen to my list of questions about who, what, when, where and why and then fill in all the details. Once done, he would then muse about certain things that he felt were important to pass on to future generations. There were certain ones he brought up every time so, clearly, they came from his heart, not just his head.

In no particular order, here they are.

Fig. 14P—A member of King Neptune's royal family pays Captain Seiberlich a visit on his bridge as *Hornet* crosses the equator during the *Apollo 11* mission. There was a high degree of mutual respect between Commanding Officer and crew.

Hornet Crew

Rear Admiral Seiberlich was extremely proud of the performance of *Hornet's* crew during the *Apollo 11* recovery. He was well aware they had just returned from a grueling seven month tour of combat duty off the coast of Vietnam. Shipboard life on an aircraft carrier at sea is hard during the best of times given the noise and danger of constant aircraft launchings and landings. A combat situation in the hot, humid tropics where a serious incident could occur at any moment makes it even worse.

Just three weeks after returning from this WestPac cruise, and one week after Seiberlich was appointed Commanding Officer, *Hornet* was selected for the *Apollo 11* mission. Long scheduled family plans and personnel leaves were cancelled and many had to report back aboard to begin the work-up for the recovery mission. They had to put up with a severely altered ship-operating environment. The Secret Service created additional turmoil by doing personnel background checks and sealing off random compartments. NASA layered the biological quarantine requirements over all activities with two large MQFs and other quarantine equipment taking up half of Hangar Bay #2. Several hundred civilians (aka landlubbers) were onboard, with 155 of these being press personnel roaming the ship, snapping photos and filming B-roll footage at any moment. Even so, the roughly 2,000 crewmen knew their duty and pulled together as a team, the ultimate complement for any ship's crew.

B&A Crane Operator

Seiberlich was inspired by the fact a nineteen-year-old bosun's mate operated the B&A crane that hauled *Columbia* onboard with its precious load of Moon rocks. For a few minutes, the pride and prestige of the entire nation—as well as $25 billion invested in the Apollo program up to that time—rested in the hands of an enlisted man who had some training, but no prior experience in this activity. The Admiral recalled how the fate of democracy rested in the hands of nineteen-year old soldiers and sailors in WWII who rose to the challenge in spite of incredible hardships. A generation later, similar young men answered the call of their nation. He felt America's democratic underpinnings enabled, if not encouraged, ordinary citizens to take initiative when required and to respond enthusiastically to difficult situations.

President Nixon and the Deckhand

Another insightful moment of reflection for Seiberlich is an incident mentioned in Chapter 11 when President Nixon was leaving the ship. After giving his welcoming speech to the astronauts, Nixon went up to the flight deck to board his helicopter. A number of *Hornet* crewmen had gathered on the teak-covered deck, hoping to catch a glimpse of the President before his departure. As Nixon approached a group of enlisted sailors, Seiberlich had some concerns. Not only had many of these sailors effectively been away from home for eight months, much of that time was spent in a combat environment. Also, all branches of the military had experienced some racial tensions during the Vietnam War, adding an element of stress on Navy ships.

One of the first individuals Nixon walked up to was an African-American aircraft refueling deckhand, wearing his purple-shirt. The President shook his hand and said "Job well done, young man." Seiberlich froze for a moment, because very rarely was *any* American offered the opportunity to speak directly to the President of the United States. When given that opportunity, most people would offer up a personal agenda item of one sort or another. In this case, the man immediately answered "Yessir, we're *Hornet*." Seiberlich felt that was the most eloquent response anyone could have made, even if they prepared for it all their lives. He lobbied to have that phrase be the title of this book.

Apollo 12 Recovery Motto

Seiberlich had few regrets, but one involved the motto selected for the recovery of *Apollo 12*. His sole focus in creating the slogans for both recoveries was to establish a mind-set of "safety first" for members of the recovery team. Thus, the *Apollo 11* motto of "Hornet Plus Three" was intended to convey a message to embarked military and civilian personnel alike that everyone would safely return from the recovery operation, with the addition of three astronauts. No one was to take unusual risks unless unforeseen and dire circumstances arose.

Having achieved this level of safety during the *Apollo 11* recovery, then-Captain Seiberlich wanted to ensure a repeat performance for *Apollo 12*. His selection of the motto "Three More Like Before" was simply intended to convey the idea that if all the SIMEXs were well executed, they would have an equally successful effort for *Apollo 12*. Unfortunately, some people outside of the Navy interpreted it more like "Ho-hum, gotta do it all over again," and considered it a slight to the risks and accomplishments of the *Apollo 12* crew. Seiberlich was mortified when he learned of this, because he felt a particularly close kinship with the "All Navy Team" that flew the second lunar landing mission.

Seiberlich Legacy

RADM Carl Seiberlich felt privileged to have served his country for thirty-seven years as a naval officer. As the Commander of Primary Landing Area Recovery Group for *Apollo 11* and *Apollo 12*, he felt greatly honored at being selected to place the exclamation mark at the end of President Kennedy's challenge to accomplish a safe Moon landing within the 1960's.

> Fig 14Q (opposite)—Commendations for the recovery team poured in from many sources. Some, such as this one from Admiral John McCain (CINCPAC), clearly reflect the impact of the *Apollo 11* mission on the world geopolitical situation.

```
                    USS HORNET (CVS-12)
                    FPO San Francisco 96601

                                              20 August 1969


From:  Commanding Officer, USS HORNET (CVS12)
To:    LTJG RICHARD F. POWERS
Subj:  Commendation; delivery of

Encl:  (1) CINCPAC msg 110102Z AUG 1969
       (2) CINCPACFLT msg 120154Z AUG 1969

1. The Commander in Chief, Pacific has commended all hands,
and especially the key personnel listed in enclosure (1),
for the outstanding job they did in the recovery of the Apollo
ELEVEN Astronauts. ADM McCAIN has indicated the significance
and importance of HORNET's contribution to the overall effort
and pointed out the beneficial effect HORNET had on the accom-
plishment of the President's world wide trip. The President,
himself, has indicated his pleasure with the performance of
the ship.

2. In recognition of your accomplishments, enclosures (1) and
(2) are delivered with pleasure.

                                      C. J. SEIBERLICH

Copy to:
BUPERS
Service jacket
```

THE SECRETARY OF THE NAVY
WASHINGTON

The Secretary of the Navy takes pleasure in presenting the MERITORIOUS UNIT COMMENDATION to

MANNED SPACECRAFT RECOVERY FORCE PACIFIC (TF-130)

for service as set forth in the following

CITATION:

For meritorious service while supporting the National Aeronautics and Space Administration's APOLLO manned space-flight operations in the Pacific Area from 1 July 1967 to 26 July 1969. The Manned Spacecraft Recovery Force Pacific conducted recovery operations vital to the success of Project APOLLO in coordinated operational employment of numerous disparate Navy, Air Force, Marine and Army units and civil agencies. The painstaking planning and attention which this task force devoted to each of the eight APOLLO missions resulted in flawlessly-executed Pacific recoveries. Consistently meeting and sustaining the highest standards of excellence throughout its recovery mission, and in the development and perfection of astronaut and spacecraft recovery techniques and procedures the Manned Spacecraft Recovery Force Pacific, by its expertise, alertness, and dedication, reflected great credit upon itself and the National Aeronautics and Space Administration, and upheld the highest traditions of the United States Naval Service.

John H. Chafee
Secretary of the Navy

Fig 14R—Following the flawless recovery of *Apollo 11*, the Secretary of the Navy awarded Task Force 130 a Meritorious Unit Commendation.

Appendix B
Time-lines

Time-line of the "Space Race"

1957: U.S.S.R. launches *Sputnik 1*, the first artificial satellite to orbit the Earth, October 4.
U.S.S.R. launches *Sputnik 2*, carrying the first animal into orbit, a dog named Laika, November 3.
1958: U.S. launches its first satellite, *Explorer I*, February 1.
Nikita Khrushchev becomes premier of the Soviet Union, March 27.
NASA organization is established to coordinate U.S. space efforts, July 29.
1959: U.S. launches and recovers monkeys "Able" and "Baker," a sub-orbital flight, May 28.
Soviet probe *Luna 2* becomes the first man-made object on the Moon (crashed), September 14.
Soviet probe *Luna 3* takes the first pictures of the darkside of the Moon, October 6.
1960: U.S.S.R. *Sputnik 5* flight successfully orbits dogs "Belka" and "Strelka," August 19.
1961: John F. Kennedy inaugurated as 35th president of the United States, January 20.
NASA launches MR-2, carrying a chimpanzee named Ham on a sub-orbital flight, January 31.
Cosmonaut Valentin Bondarenko dies in a flight simulator fire, March 23.
U.S.S.R. launches *Vostok 1* with Yuri Gagarin, the first human spaceflight, one Earth orbit, April 12.
NASA launches *Freedom 7* with Alan Shepard, the first American to fly into sub-orbital space, May 5.
U.S.S.R. probe *Venera 1* does a fly-by of Venus, May 19.
President John F. Kennedy commits 100 special forces "military advisers" to South Vietnam, May 13.
President John F. Kennedy announces his intention for the U.S. to place a man on the Moon, May 25.
NASA launches *Liberty Bell 7* with Gus Grissom, a sub-orbital flight; capsule sinks during retrieval, July 21.

U.S.S.R. launches *Vostok 2* with Gherman Titov, the first human to stay in orbit a full day, August 6.

NASA launches MA-5, carrying a chimpanzee named Enos on a 2-orbit flight, November 29.

1962: NASA launches *Friendship 7* with John Glenn, the first American to orbit the Earth, February 20.

NASA launches *Aurora 7* with Scott Carpenter on a 3-orbit flight; capsule lands 402 km from EOM, May 24.

NASA launches *Sigma 7* with Wally Schirra on a 6-orbit flight, October 3.

Cuban Missile Crisis begins; U.S. Navy blockade lasts 2 weeks until Soviet missiles are removed, October 15.

1963: NASA launches *Faith 7* with Gordon Cooper, the first American to spend over a day in space, May 15.

U.S.S.R. launches *Vostok 6* with Valentina Tereshkova, the first woman in space, June 16.

President Ngo Dinh Dien of South Vietnam is assassinated during a military coup, November 2.

President John F. Kennedy is assassinated; Lyndon Johnson becomes 36th president, November 22.

U.S. Navy launches the first operational TRANSIT navigation satellite, December 5.

1964: U.S. Navy aircraft carriers begin limited bombing operations in South Vietnam, August 4.

Gulf of Tonkin naval incident; Congress authorizes increased U.S. involvement in Vietnam, August 7.

U.S.S.R. launches *Voskhod 1*; first space mission with more than 1 crewmember, October 12.

Premier Nikita Khrushchev is forced to resign; Leonid Brezhnev becomes premier, October 15

1965: U.S. Navy aircraft carriers begin sustained air operations against North Vietnam, March 2.

U.S. Marines are deployed to Vietnam, the first combat troops committed to the war, March 8.

NASA launches *Gemini 3*; first American 2-man crew in space, March 23.

U.S.S.R. mission *Voskhod 2*; Alexi Leonov performs the first human space walk, March 18.

NASA mission *Gemini 4*; Edward White performs the first American space walk, June 3.

NASA missions *Gemini 6A* and *Gemini 7* accomplish the first American space rendezvous, December 15.

1966: Soviet probe *Luna 9*; first spacecraft to make a soft landing on the Moon and transmit photos, February 3.

Soviet probe *Venera 3* crash-lands on Venus, March 1.

NASA mission *Gemini 8* makes the first orbital docking and emergency landing of a spacecraft, March 17.

NASA mission *Gemini 10* performs rendezvous & docking with spacecraft and 2 EVAs, July 18-21.

NASA mission *Gemini 11* achieves altitude record and performs 2 EVAs, September 12-15.

NASA mission *Gemini 12* sets EVA record with 5.5 total hours over 3 EVAs, November 11-15.

1967: NASA mission *Apollo 1* catches fire during a launch pad test, killing all 3 crewmembers, January 27.

U.S.S.R. mission *Soyuz 1*; Vladmir Komarov becomes the first spaceflight death, April 24.

USS *Liberty* is attacked and severely damaged by Israeli forces in the Mediterranean Sea, June 8

1968: USS *Pueblo* is attacked and captured by North Korean forces in international waters, January 23.

North Vietnamese and Viet Cong soldiers launch the Tet Offensive in South Vietnam, January 31.

Yuri Gagarin, now deputy training director at Russia's Star City, is killed during a training incident, March 27.

President Johnson announces he will not seek re-election, March 31.

NASA launches *Apollo 7*; first American 3-man crew in space, October 11.

NASA mission *Apollo 8*; first humans to orbit a celestial body other than Earth, December 24.

1969: U.S.S.R. missions *Soyuz 4* and Soyuz 5 dock and exchange crews via EVA, January 16.

Richard M. Nixon inaugurated as 37th president of the United States, January 20.

NASA Mission *Apollo 10* orbits the Moon and tests the lunar landing module, May 18.

President Nixon announces a significant reduction of U.S. troops in South Vietnam, June 8.

U.S.S.R. N-1 rocket explodes on the launch pad, eliminating any chance of a Moon landing, July 3.

NASA Mission *Apollo 11* lands the first 2 humans to ever walk on the Moon, July 20.

President Nixon visits U.S. troops during his only trip to South Vietnam, July 30.

Time-line of *Apollo 11* Recovery

May 15, 1969:	USS *Hornet* returns to Long Beach from a third and final Vietnam deployment.
May 25:	Captain Carl Seiberlich joins *Hornet* as commanding officer.
June 1:	*Hornet* is nominated as Primary Recovery Ship (PRS) for *Apollo 11*
June 5:	*Hornet* is designated as PRS for *Apollo 11*.
June 12-25:	*Hornet* commences planning stage of recovery mission. She is outfitted with significant specialized electronic communication equipment and embarks 130 technicians from Navy, NASA, GE, ABC, Mutual Radio, Voice of America and others to provide support.
June 26-July 2:	*Hornet* sails to San Diego where she embarks HS-4 with eight SH-3D Sea King helicopters, VAW-111 with four E-1B Tracer early warning aircraft, VR-30 with three C-1A Trader CODs and their support equipment. These recovery units comprise over 250 men. *Hornet* sails to Pearl Harbor and is "chopped" to TF-130 upon arrival.
July 3-6:	Much NASA and Navy support equipment is onloaded, including a NASA "boilerplate" training capsule. The UDT team embarks with their specialized equipment.
July 7-9:	*Hornet* performs numerous SIMEXs (Simulated Recovery Exercises) in Hawaiian waters to verify basic recovery procedures and test communications equipment.

July 10-11:		Final load-on of supplies in Pearl Harbor, including two NASA MQFs and much quarantine equipment. The majority of the TV, radio and press personnel come aboard.
July 12-16:		*Hornet* sails to her Primary Launch Abort position, 1,650 miles southwest of Hawaii. SIMEXs conducted along the way plus equator crossing ceremony held on July 15. *Hornet* position for *Apollo 11* Launch and TLI Abort on July 16: latitude 3°-00', longitude 165° 00'W.
July 17-22:		After a successful spacecraft launch and TLI burn, *Hornet* sails northwest toward the End-Of-Mission (EOM) target area, 1,200 miles southwest of Pearl Harbor. She sails slowly along the Mid-Pacific Line for deep space abort in case a problem arose with the *Apollo 11* flight. Crew continues all weather, day and evening SIMEX training. By July 22, a total of 16 SIMEX rehearsals are completed.
July 22:		A serious weather system approaches the planned recovery area, causing NASA to relocate the End-Of-Mission site 250 miles to the northeast. *Hornet* position at planned splashdown area on July 23—latitude 10°-56'N, longitude 172°-24'W.
July 23-24:		*Hornet* steams at flank speed to the new splashdown site, 920 miles southwest of Pearl Harbor. *Hornet* COD flies to Johnston Island to pick up ADM John McCain (CINPAC). *Hornet* position for *Apollo 11* splashdown and recovery on July 24: latitude 13°-25'N, longitude 168°-58'W (12 nautical miles downrange and upwind from the EOM target point).
July 24	0400:	Two hours from the scheduled splashdown, television crews prepare to broadcast the recovery and ceremonies to the second largest worldwide television audience.
	0418:	*Hornet* launches aircraft—5 SH3D's and 2 E1B's.
	0504:	*Hawaii Rescue One* (ARRS HC-130H) reports "on station."
	0505:	*Hawaii Rescue Two* (ARRS HC-130H) reports "on station."
	0512:	President Nixon and party arrive onboard.
	0535:	*Apollo 11* atmospheric re-entry and blackout begins.
	0539:	*Apollo 11* fireball spotted by *Hawaii Rescue One*.
	0542:	Ship has radar contact at 65nm; spacecraft drogue parachutes deploy.
	0545:	Spacecraft main parachutes deploy.
	0546:	VHF voice & recovery beacon contact.
	0550:	*Columbia* splashed down, 13 miles downwind of *Hornet*, in Stable 2 position.
	0556:	CM in Stable 1 position.
	0558:	First UDT swimmer in the water.
	0627:	First astronaut exits CM into raft.

0644: Astronaut decontamination process complete.
0649: First astronaut hoisted into *Recovery* Helicopter #66.
0653: Astronauts arrive onboard *Hornet*.
0658: Astronauts enter the MQF.
0755: President Nixon greets the astronauts; Chaplain Piirto says a prayer.
0811: President Nixon departs for Johnston Island.
0812: Admiral McCain addresses the ship's crew.
0849: *Columbia* is hoisted aboard by the B&A crane.
0858: CM secured in HB2 and connected to MQF via a plastic tunnel.
0915: *Hornet* heads for Pearl Harbor
0920: HC-130H STAR pickup of news film from flight deck

July 25: Lunar samples, film and personal recording containers removed from the CM are processed and packaged in the MQF and then launched on two COD planes, each carrying half of the 46 pounds of Moon rocks. In case of accident, one flies to Johnston Island, the other to Hickam AFB in Honolulu. From there, the sample containers are flown to NASA's Lunar Receiving Laboratory in Houston. *Hornet's* post office continues to cancel over 248,000 letters and cards with her recovery seal and date of recovery.

July 26: *Hornet* arrives in Pearl Harbor with a broomstick attached to her mast. Spacecraft *Columbia* and ship's crew man the rails. The MQF (with astronauts inside) is off-loaded, brief welcoming ceremonies are held. The trailer travels by motorcade to nearby Hickam AFB, is installed in a USAF C-141 cargo plane and flown to Houston's Johnson Space Center where the astronauts undergo two more weeks of quarantine and debriefing. *Columbia* is also off-loaded, deactivated, and flown aboard a USAF C-133 from Hickam AFB to Ellington AFB for technical review.

July 28-August 1: *Hornet* sails for Long Beach following a very successful and historic mission.

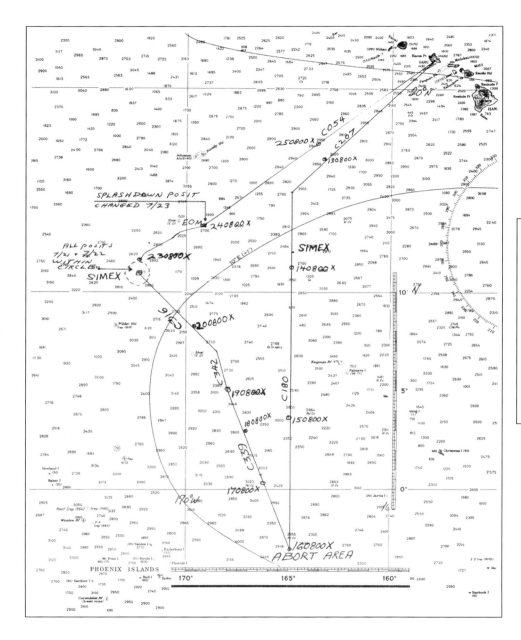

Fig 15A—The navigation chart from the *U.S.S. Hornet Apollo 11 Recovery Mission* shows the lengthy voyage and numerous SIMEX activities undertaken by the ship. At 8 a.m. on July 23, *Hornet* was loitering at the original EOM point (the sunrise "position fix" labelled 230800X). The erratic hand-drawn line that starts at this point and ends at 240800X reflects the overnight high-speed dash made by the ship when the planned splashdown point was moved due to thunderstorms in the original area. Most of this cruise segment was based on "dead reckoning" techniques. *Hornet's* navigation team was finally able to get an accurate position fix using their sextants just an hour before splashdown.

Appendix C
Key Speeches

President John F. Kennedy's speech before Congress, May 25, 1961

Mr. Speaker, Mr. Vice President, my co-partners in Government, gentlemen and ladies:

The Constitution imposes upon me the obligation to "from time to time give to the Congress information of the State of the Union." While this has traditionally been interpreted as an annual affair, this tradition has been broken in extraordinary times.

These are extraordinary times. And we face an extraordinary challenge. Our strength as well as our convictions have imposed upon this nation the role of leader in freedom's cause. No role in history could be more difficult or more important. We stand for freedom. That is our conviction for ourselves-that is our only commitment to others. No friend, no neutral and no adversary should think otherwise. We are not against any man—or any nation—or any system—except as it is hostile to freedom. Nor am I here to present a new military doctrine, bearing any one name or aimed at any one area. I am here to promote the freedom doctrine . . . .

. . . Finally, if we are to win the battle that is now going on around the world between freedom and tyranny, the dramatic achievements in space which occurred in recent weeks should have made clear to us all, as did the Sputnik in 1957, the impact of this adventure on the minds of men everywhere, who are attempting to make a determination of which road they should take. Since early in my term, our efforts in space have been under review. With the advice of the Vice President, who is Chairman of the National Space Council, we have examined where we are strong and where we are not, where we may succeed and where we may not. Now it is time to take longer strides—time for a great new American enterprise—time for this nation to take a clearly leading role in space achievement, which in many ways may hold the key to our future on Earth.

I believe we possess all the resources and talents necessary. But the facts of the matter are that we have never made the national decisions or marshaled the national resources required for such leadership. We have never specified long-range goals on an urgent time schedule, or managed our resources and our time so as to insure their fulfillment.

Recognizing the head-start obtained by the Soviets with their large rocket engines, which gives them many months of lead-time, and recognizing the likelihood that they will exploit this lead for some time to come in still more impressive successes, we nevertheless are required to make new efforts on our own. For while we cannot guarantee that we shall one day be first, we can guarantee that any failure to make this effort will make us last. We take an additional risk by making it in full view of the world, but as shown by the feat of astronaut Shepard, this very risk enhances our stature when we are successful. But this is not merely a race. Space is open to us now; and our eagerness to share its meaning is not governed by the efforts of others. We go into space because whatever mankind must undertake, free men must fully share.

I therefore ask the Congress, above and beyond the increases I have earlier requested for space activities, to provide the funds which are needed to meet the following national goals:

First, I believe that this nation should commit itself to achieving the goal, before this decade is out, of landing a man on the Moon and returning him safely to the earth. No single space project in this period will be more impressive to mankind, or more important for the long-range exploration of space; and none will be so difficult or expensive to accomplish. We propose to accelerate the development of the appropriate lunar space craft. We propose to develop alternate liquid and solid fuel boosters, much larger than any now being developed, until certain which is superior.

We propose additional funds for other engine development and for unmanned explorations—explorations which are particularly important for one purpose which this nation will never overlook: the survival of the man who first makes this daring flight. But in a very real sense, it will not be one man going to the Moon—if we make this judgment affirmatively, it will be an entire nation. For all of us must work to put him there.

Secondly, an additional twenty-three million dollars, together with seven million dollars already available, will accelerate development of the Rover nuclear rocket. This gives promise of some day providing a means for even more exciting and ambitious exploration of space, perhaps beyond the Moon, perhaps to the very end of the solar system itself.

Third, an additional fifty million dollars will make the most of our present leadership, by accelerating the use of space satellites for world-wide communications.

Fourth, an additional seventy-five million dollars—of which fifty-three million dollars is for the Weather Bureau—will help give us at the earliest possible time a satellite system for world-wide weather observation.

Let it be clear—and this is a judgment which the Members of the Congress must finally make—let if be clear that I am asking the Congress and the country to accept a firm commitment to a new course of action—a course which will last for many years and carry very heavy costs: 531 million dollars in fiscal '62—an estimated seven to nine billion dollars additional over the next five years. If we are to go only half way, or reduce our sights in the face of difficulty, in my judgment it would be better not to go at all.

Now this is a choice which this country must make, and I am confident that under the leadership of the Space Committees of the Congress, and the Appropriating Committees, that you will consider the matter carefully.

It is a most important decision that we make as a nation. But all of you have lived through the last four years and have seen the significance of space and the adventures in space, and no one can predict with certainty what the ultimate meaning will be of mastery of space.

I believe we should go to the Moon. But I think every citizen of this country as well as the Members of the Congress should consider the matter carefully in making their judgment, to which we have given attention over many weeks and months, because it is a heavy burden, and there is no sense in agreeing or desiring that the United States take an affirmative position in outer space, unless we are prepared to do the work and bear the burdens to make it successful. If we are not, we should decide today and this year.

This decision demands a major national commitment of scientific and technical manpower, materiel and facilities, and the possibility of their diversion from other important activities where they are already thinly spread. It means a degree of dedication, organization and discipline which have not always characterized our research and development efforts. It means we cannot afford undue work stoppages, inflated costs of material or talent, wasteful interagency rivalries, or a high turnover of key personnel.

New objectives and new money cannot solve these problems. They could in fact, aggravate them further—unless every scientist, every engineer, every serviceman, every technician, contractor, and civil servant gives his personal pledge that this nation will move forward, with the full speed of freedom, in the exciting adventure of space.

President John F. Kennedy's speech at Rice University in Houston, Texas, on September 12, 1962.

President Pitzer, Mr. Vice President, Governor, Congressman Thomas, Senator Wiley, and Congressman Miller, Mr. Webb, Mr. Bell, scientists, distinguished guests, and ladies and gentlemen:

I appreciate your president having made me an honorary visiting professor, and I will assure you that my first lecture will be very brief.

I am delighted to be here and I'm particularly delighted to be here on this occasion.

We meet at a college noted for knowledge, in a city noted for progress, in a state noted for strength, and we stand in need of all three, for we meet in an hour of change and challenge, in a decade of hope and fear, in an age of both knowledge and ignorance. The greater our knowledge increases, the greater our ignorance unfolds.

Despite the striking fact that most of the scientists that the world has ever known are alive and working today, despite the fact that this Nation's own scientific manpower is doubling every twelve years in a rate of growth more than three times that of our population as a whole, despite that, the vast stretches of the unknown and the unanswered and the unfinished still far outstrip our collective comprehension.

No man can fully grasp how far and how fast we have come, but condense, if you will, the 50,000 years of man's recorded history in a time span of but a half-century. Stated in these terms, we know very little about the first forty years, except at the end of them advanced man had learned to use the skins of animals to cover them. Then about ten years ago, under this standard, man emerged from his caves to construct other kinds of shelter. Only five years ago man learned to write and use a cart with wheels. Christianity began less than two years ago. The printing press came this year, and then less than two months ago, during this whole fifty-year span of human history, the steam engine provided a new source of power. Newton explored the meaning of gravity. Last month electric lights and telephones and automobiles and airplanes became available. Only last week did we develop penicillin and television and nuclear power, and now if America's new spacecraft succeeds in reaching Venus, we will have literally reached the stars before midnight tonight.

This is a breathtaking pace, and such a pace cannot help but create new ills as it dispels old, new ignorance, new problems, new dangers. Surely the opening vistas of space promise high costs and hardships, as well as high reward.

So it is not surprising that some would have us stay where we are a little longer to rest, to wait. But this city of Houston, this state of Texas, this country of the United States was not built by those who waited and rested and wished to look behind them. This country was conquered by those who moved forward—and so will space.

William Bradford, speaking in 1630 of the founding of the Plymouth Bay Colony, said that all great and honorable actions are accompanied with great difficulties, and both must be enterprised and overcome with answerable courage.

If this capsule history of our progress teaches us anything, it is that man, in his quest for knowledge and progress, is determined and cannot be deterred. The exploration of space will go ahead, whether we join in it or not, and it is one of the great adventures of all time, and no nation which expects to be the leader of other nations can expect to stay behind in this race for space.

Those who came before us made certain that this country rode the first waves of the industrial revolution, the first waves of modern invention, and the first wave of nuclear power, and this generation does not intend to founder in the backwash of the coming age of space. We mean to be a part of it—we mean to lead it. For the eyes of the world now look into space, to the Moon and to the planets beyond, and we have vowed that we shall not see it governed by a hostile flag of conquest, but by a banner of freedom and peace. We have vowed that we shall not see space filled with weapons of mass destruction, but with instruments of knowledge and understanding.

Yet the vows of this Nation can only be fulfilled if we in this Nation are first, and, therefore, we intend to be first. In short, our leadership in science and industry, our hopes for peace and security, our obligations to ourselves as well as others, all require us to make this effort, to solve these mysteries, to solve them for the good of all men, and to become the world's leading space-faring nation.

We set sail on this new sea because there is new knowledge to be gained, and new rights to be won, and they must be won and used for the progress of all people. For space science, like nuclear science and all technology, has no conscience of its own. Whether it will become a force for good or ill depends on man, and only if the United States occupies a position of pre-eminence can we help decide whether this new ocean will be a sea of peace or a new terrifying theater of war. I do not say that we should or will go unprotected against the hostile misuse of space any more than we go unprotected against the hostile use of land or sea, but I do say that space can be explored and mastered without feeding the fires of war, without repeating the mistakes that man has made in extending his writ around this globe of ours.

There is no strife, no prejudice, no national conflict in outer space as yet. Its hazards are hostile to us all. Its conquest deserves the best of all mankind, and its opportunity for peaceful cooperation many never come again. But why, some say, the Moon? Why choose this as our goal? And they may well ask why climb the highest mountain? Why, thirty-five years ago, fly the Atlantic? Why does Rice play Texas?

We choose to go to the Moon. We choose to go to the Moon in this decade and do the other things, not because they are easy, but because they are hard, because that goal will serve to organize and measure the best of our energies and skills, because that challenge is one that we are willing to accept, one we are unwilling to postpone, and one which we intend to win, and the others, too.

It is for these reasons that I regard the decision last year to shift our efforts in space from low to high gear as among the most important decisions that will be made during my incumbency in the office of the Presidency.

In the last twenty-four hours we have seen facilities now being created for the greatest and most complex exploration in man's history. We have felt the ground shake and the air shattered by the testing of a Saturn C-1 booster rocket, many times as powerful as the Atlas which launched John Glenn, generating power equivalent to 10,000 automobiles with their accelerators on the floor. We have seen the site where five F-1 rocket engines, each one as powerful as all eight engines of the Saturn combined, will be clustered together to make the advanced Saturn missile, assembled in a new building to be built at Cape Canaveral as tall as a forty-eight story structure, as wide as a city block, and as long as two lengths of this field.

Within these last nineteen months at least forty-five satellites have circled the earth. Some forty of them were made in the United States of America and they were far more sophisticated and supplied far more knowledge to the people of the world than those of the Soviet Union.

The Mariner spacecraft now on its way to Venus is the most intricate instrument in the history of space science. The accuracy of that shot is comparable to firing a missile from Cape Canaveral and dropping it in this stadium between the forty-yard lines.

Transit satellites are helping our ships at sea to steer a safer course. Tiros satellites have given us unprecedented warnings of hurricanes and storms, and will do the same for forest fires and icebergs.

We have had our failures, but so have others, even if they do not admit them. And they may be less public.

To be sure, we are behind, and will be behind for some time in manned flight. But we do not intend to stay behind, and in this decade, we shall make up and move ahead.

The growth of our science and education will be enriched by new knowledge of our universe and environment, by new techniques of learning and mapping and observation, by new tools and computers for industry, medicine, the home as well as the school. Technical institutions, such as Rice, will reap the harvest of these gains.

And finally, the space effort itself, while still in its infancy, has already created a great number of new companies, and tens of thousands of new jobs. Space and related industries are generating new demands in investment and skilled personnel, and this city and this state, and this region, will share greatly in this growth. What was once the furthest outpost on the old frontier of the West will be the furthest outpost on the new frontier of science and space. Houston, your city of Houston, with its Manned Spacecraft Center, will become the heart of a large scientific and engineering community. During the next five years, the National Aeronautics and Space Administration expects to double the number of scientists and engineers in this area, to increase its outlays for salaries and expenses to $60 million a year; to invest some $200 million in plant and laboratory facilities; and to direct or contract for new space efforts over $1 billion from this center in this city.

To be sure, all this costs us all a good deal of money. This year's space budget is three times what it was in January 1961, and it is greater than the space budget of the previous eight years combined. That budget now stands at $5,400 million a year—a staggering sum, though somewhat less than we pay for cigarettes and cigars every year. Space expenditures will soon rise some more, from 40 cents per person per week to more than fifty cents a week for every man, woman and child in the United States, for we have given this program a high national priority—even though I realize that this is in some measure an act of faith and vision, for we do not now know what benefits await us. But if I were to say, my fellow citizens, that we shall send to the Moon, 240,000 miles away from the control station in Houston, a giant rocket more than 300 feet tall, the length of this football field, made of new metal alloys, some of which have not yet been invented, capable of standing heat and stresses several times more than have ever been experienced, fitted together with a precision better than the finest watch, carrying all the equipment needed for propulsion, guidance, control, communications, food and survival, on an untried mission, to an unknown celestial body, and then return it safely to earth, re-entering the atmosphere at speeds of over 25,000 miles per hour, causing heat about half that of the temperature of the sun—almost as hot as it is here today—and do all this, and do it right, and do it first before this decade is out—then we must be bold.

I'm the one who is doing all the work, so we just want you to stay cool for a minute. [laughter]

However, I think we're going to do it, and I think that we must pay what needs to be paid. I don't think we ought to waste any money, but I think we ought to do the job. And this will be done in the decade of the Sixties. It may be done while some of you are still here at school at this college and university. It will be done during the terms of office of some of the people who sit here on this platform. But it will be done. And it will be done before the end of this decade.

And I am delighted that this university is playing a part in putting a man on the Moon as part of a great national effort of the United States of America.

Many years ago the great British explorer George Mallory, who was to die on Mount Everest, was asked why did he want to climb it. He said, "Because it is there."

Well, space is there, and we're going to climb it, and the Moon and the planets are there, and new hopes for knowledge and peace are there. And, therefore, as we set sail we ask God's blessing on the most hazardous and dangerous and greatest adventure on which man has ever embarked.

President Richard M. Nixon's Telephone Conversation with the Astronauts on the Moon, July 20, 1969

President Richard M. Nixon: Hello Neil and Buzz, I am talking to you by telephone from the Oval Room at the White House, and this certainly has to be the most historic telephone call ever made from the White House.

I just can't tell you how proud we all are of what you have done. For every American this has to be the proudest day of our lives, and for people all over the world I am sure that they, too, join with Americans in recognizing what an immense feat this is.

Because of what you have done, the heavens have become a part of man's world, and as you talk to us from the Sea of Tranquility, it inspires us to redouble our efforts to bring peace and tranquility to Earth.

For one priceless moment in the whole history of man all the people on this Earth are truly one—one in their pride in what you have done and one in our prayers that you will return safely to Earth.

Astronaut Armstrong: Thank you, Mr. President. It is a great honor and privilege for us to be here representing not only the United States, but men of peaceable nations, men with an interest and a curiosity, and men with a vision for the future. It is an honor for us to be able to participate here today.

The President: Thank you very much, and I look forward, all of us look forward, to seeing you on the *Hornet* on Thursday.

Astronaut Armstrong: Thank you. We look forward to that very much.

President Richard M. Nixon's Prepared Speech, In the event of an *Apollo 11* Disaster (written by William Safire)

NASA and the Nixon administration had to be prepared in case something went seriously wrong with the *Apollo 11* mission. One such contingency would have been the failure of the Lunar Module engine to ignite and place astronauts Neil Armstrong and Buzz Aldrin back in orbit around the Moon to rendezvous with the CSM for the return to Earth.

The following is the text of an unused speech that President Nixon would have given as the astronauts lived out their final hours on the Moon's surface:

Fate has ordained that the men who went to the Moon to explore in peace will stay on the Moon to rest in peace.

These brave men, Neil Armstrong and Edwin Aldrin, know that there is no hope for their recovery. But they also know that there is hope for mankind in their sacrifice.

These two men are laying down their lives in mankind's most noble goal: the search for truth and understanding.

They will be mourned by their families and friends; they will be mourned by their nation; they will be mourned by the people of the world; they will be mourned by a Mother Earth that dared send two of her sons into the unknown.

In their exploration, they stirred the people of the world to feel as one; in their sacrifice, they bind more tightly the brotherhood of man.

In ancient days, men looked at stars and saw their heroes in the constellations.

In modern times, we do much the same, but our heroes are epic men of flesh and blood.

Others will follow, and surely find their way home. Man's search will not be denied.

But these men were the first, and they will remain the foremost in our hearts.

For every human being who looks up at the Moon in the nights to come will know that there is some corner of another world that is forever mankind.

Needless to say, President Nixon was greatly relieved as he watched the retrieval helicopter pluck the three astronauts from the ocean and bring them safely back to the USS *Hornet*.

President Richard M. Nixon
Congratulates the Returned Astronauts Aboard the *Hornet*, July 24, 1969

Neil, Buzz, and Mike:

I want you to know that I think I am the luckiest man in the world, and I say this not only because I have the honor to be President of the United States, but particularly because I have the privilege of speaking for so many in welcoming you back to Earth.

I can tell you about all the messages we have received in Washington. Over 100 foreign governments, emperors, presidents, prime ministers, and kings, have sent the most warm messages that we have ever received. They represent over 2 billion people on this Earth, all of them who have had the opportunity, through television, to see what you have done.

Then I also bring you messages from members of the Cabinet and Members of the Senate, Members of the House, the space agency, from the streets of San Francisco where people stopped me a few days ago, and you all love that city, I know, as I do.

But most important, I had a telephone call yesterday. The toll wasn't, incidentally, as great as the one I made to you fellows on the Moon. I made that collect, incidentally, in case you didn't know. But I called three, in my view, three of the greatest ladies and most courageous ladies in the whole world today—your wives.

From Jan, Joan and Pat, I bring their love and their congratulations. We think it is just wonderful that they could participate at least by television in this return. We are only sorry they couldn't be here.

Also, I will let you in on a little secret. I made a date with them. I invited them to dinner on the thirteenth of August, right after you come out of quarantine. It will be a state dinner held in Los Angeles. The Governors of all the fifty States will be there, the Ambassadors, others from around the world and in America. They told me that you would come, too. All I want to know is: Will you come? We want to honor you then.

Astronaut Neil Armstrong: We will do anything you say, Mr. President, anytime.

The President: One question, I think all of us would like to ask: As we saw you bouncing around in that boat out there, I wonder if that wasn't the hardest part of the journey. Did any of you get seasick?

Astronaut Armstrong: No, we didn't, and it was one of the harder parts, but it was one of the most pleasant, we can assure you.

The President: Well, I just know that you can sense what we all sense. When you get back now incidentally, have you been able to follow some of the things that happened since you have been gone? Did you know about the All-Star Game?

Astronaut Buzz Aldrin: Yes, sir. The capsule communicators have been giving us daily reports.

The President: Were you American League or National League?

Astronaut Armstrong: I'm a National League man.

Astronaut Aldrin: I'm non-partisan, sir.

The President: There is the politician in the group. (Pointing toward Aldrin.)

Astronaut Armstrong: We're sorry you missed that game.

The President: Oh, you knew that too?

Astronaut Armstrong: We heard about the rain. We haven't learned to control the weather yet, but that is something we can look forward to as tomorrow's challenge.

The President: Well, I can only summarize it because I don't want to hold you now. You have so much more to do. Gee, you look great. Do you feel as great as you look?

Astronaut Armstrong: We feel great.

Astronaut Collins: We feel just perfect, Mr. President.

The President: Frank Borman says you are a little younger by reason of having gone into space. Is that right? Do you feel a little bit younger?

Astronaut Collins: We're a lot younger than Frank Borman!

The President: He is over there. Come on over, Frank, so they can see you. Are you going to take that lying down?

Astronaut Collins: It looks like he has aged in the last couple of weeks.

Colonel Frank Borman: They look a little heavy. Mr. President, the one thing I wanted—you know, we have a poet in Mike Collins. He really gave me a hard time for describing things with the words "fantastic" and "beautiful." I counted them. In three minutes up there, you used four "fantastics" and two "beautifuls."

The President: Well, just let me close off with this one thing: I was thinking, as you know, as you came down, and we knew it was a success, and it had only been eight days, just a week, a long week, that this is the greatest week in the history of the world since the Creation, because as a result of what happened in this week, the world is bigger, infinitely, and also, as I am going to find on this trip around the world, and as Secretary Rogers will find as he covers the other countries in Asia, as a result of what you have done, the world has never been closer together before. We just thank you for that. I only hope that all of us in Government, all of us in America, that as a result of what you have done, can do our job a little better.

We can reach for the stars just as you have reached so far for the stars.

We don't want to hold you any longer. Anybody have a last—how about promotions? Do you think we can arrange something?

Astronaut Armstrong: We are just pleased to be back and very honored that you were so kind as to come out here and welcome us back. We look forward to getting out of this quarantine and talking without having glass between us.

The President: Incidentally, the speeches that you have to make at this dinner can be very short. If you want to say "fantastic" or "beautiful," that is all right with us. Don't try to think of new adjectives. They have all been said.

Now, I think incidentally that all of us, the millions who are seeing us on television now, seeing you, would feel as I do, that, in a sense, our prayers have been answered, and I think it would be very appropriate if Chaplain Piirto, the Chaplain of this ship, were to offer a prayer of thanksgiving. If he would step up now.

Appendix D
Key Crew Assignments

USS Hornet	NASA
CO – CAPT Carl Seiberlich	Operations Director – Don Stullken
XO – CDR Chris Lamb	Quarantine Manager – John Stonesifer
Air Boss – CDR Irvin Patch	MQF Doctor – Dr. William Carpentier
Ops Officer – LCDR John McNally	MQF Technician – John Hirasaki
CM Retrieval team – LCDR Dick Knapp	
President Nixon EA – LTJG Dick Powers	
PAO – LTJG Tim Wilson	
Helmsman – QM2 Ken Hoback	

HS-4 & UDT	
Call Sign – *Recovery One* (helo #66)	Call Sign – *Swim Two* (helo #64)
Pilot - CDR Don Jones	Pilot – LT Richard Barrett
Copilot – LTJG Bruce Johnson	Copilot – George Conn
Crew – AWHC Norvel Wood	Crew - AWH2 Seaton
Crew – AWHC Stanley Robnett	Crew - AWH2 Curtis Hill
Flight Surgeon – Dr. Wm Carpentier	UDT - LTJG Wes Chesser
UDT (decon) – LT Clancy Hatleberg	UDT - QM3 Michael Mallory
	UDT - SN John Wolfram

HS-4 & UDT

Call Sign - *Photo* (helo #70)
Pilot – LCDR Andy Patrick
Copilot – LTJG Larry Duncan
Crew - AWH3 Gregory Benton
HS-4 photo - PH2 Milt Putnam
NASA photo – Lee Jones

Call Sign – *Swim Three* (standby helo #60)
Pilot – LT Paul Dalpiaz
Copilot – LTJG Charles Whitten
Crew - AWH2 Abram Dominguez
Crew - AWH2 Michael Miller
UDT – ENS Bob Rohrbach
UDT – GMG3 Charles Free
UDT - ADJ3 Joseph Via

Call Sign – *Swim One* (helo #53)
Pilot – LCDR Donald Richmond
Copilot – LT William Strawn
Crew - AWH3 George Klepak
Crew - AXAN James Johnson
UDT - LTJG John McLaughlin
UDT - PH2 Terry Muehlenbach
UDT - ADJ3 Mitchell Bucklew

Backups
Alternate decon swimmer: EN2 Bennett
Alternate swimmer: HM1 Holmes.

Appendix E
Photography Related Information

Determining what person should be credited with capturing every photograph taken during the *Apollo 11* recovery is an exceedingly difficult, and probably impossible, task for a number of reasons:

As far as the official Navy and news media were concerned, the shipboard environment was considered a "pool reporting" situation, since only a hundred or so of the thousands of media outlet reporters and photographers could be accommodated. As a pool operation, the film from various official news photographers was often mixed in *Hornet*'s photo lab while being processed. Often, whatever agency put a photo "on the wire" first ended up with the credit.

Also, Navy photographers were rarely credited with their photos during this timeframe. Often they were labeled with just the ship name or "Official U.S. Navy Photograph."

On this particularly historic occasion, the situation was greatly complicated by the hundreds of sailors and civilians carrying their own cameras. Over time, many of these made their way into newspaper articles, naval archives or various online websites, often losing their attribution along the way.

NASA has made a few recovery images from each manned space mission publicly accessible via online databases. However, these are marked "attribution to NASA", which really means it is available from a NASA website, not necessarily taken by a NASA photographer. Many of the photos in this book were provided courtesy of NASA, either taken from one of their websites or sent by personnel in the Media Resource Center at Johnson Space Center.

For the recovery of *Apollo 11*, there are some *general* guidelines regarding who probably captured the shot. Only two photographers were in the P*hoto* helicopter at the splashdown scene, Lee Jones and Milt Putnam. Lee was a NASA videographer and used a 16mm 70 KRM motion picture camera with a 100-foot load of film. The camera had a rotating turret for three lenses: a wide-angle lens, a medium length lens and a telephoto lens. Navy photojournalist Milt Putnam used a couple of still cameras, shooting both in black & white and color. Any film or photo that shows a close-up of the astronaut recovery taken from the air was captured by one of these two men.

FIG 18A—During *Hornet's* voyage back to Pearl Harbor, photographers from various military units and civilian organizations gathered for a group photo in front of *Columbia* in Hangar Bay #2.

However, also at that scene with a camera was one of the UDT swimmers, who took several still photographs during the helicopter hoist operation. The photographs are rarely seen since water droplets splashed on his lens and created water marks on the film. These photos were taken from ground level.

On *Hornet*, there were literally hundreds of professional and amateur photographers clicking away. The ship's photographic department was managed by PHCS Robert Lawson, an excellent photographer in his own right. While Lee and Milt were in position to capture important moments on the ship, the majority of these photos came from the ship's photographers using professional cameras. Some crewmen used personal cameras to capture candid shots of varying value. Many of these found their way into the *Hornet Cruise Book* or were kept in private storage until unearthed in support of this book. Undoubtedly, there are thousands more held in dusty scrapbooks and cardboard boxes in many attics.

The main body of recovery photographs remains unaccounted for. Following the *Apollo 11* recovery, most were shipped to the Navy Photographic Center in Washington, D.C., while others were sent to NAS North Island in San Diego. Many years after that, storage containers were sent on to the National Archives, but very few Apollo recovery photos have appeared in their files. Robert Lawson and Milt Putnam deserve specific recognition for preserving many important recovery photos and making them available to researchers.

Photo Sources

Chapter 1

Fig 1A	Pacific Recovery map (Wikipedia)
Fig 1B	NASA insignia (NASA)
Fig 1C	DDMS Organization chart (NASA)
Fig 1D	TF-140 Logo (patch scan)
Fig 1E	TF-130 Logo (patch scan)
Fig 1F	AS-202 *Hornet* (USN)

Chapter 2

Fig 2A	MR-3 Recovery (NASA S88-31378)
Fig 2B	MR-3 Recovery (NASA S61-19882)
Fig 2C	MR-4 Recovery (NASA S61-02826)
Fig 2D	MA-9 Recovery UDT (NASA MSFC-6413216)
Fig 2E	MA-9 Recovery Capsule (NASA S63-07854)
Fig 2F	*Gemini 4* Recovery (NASA S65-34044)
Fig 2G	*Gemini 5* Recovery (NASA S65-46630)
Fig 2H	*Gemini 8* Recovery (NASA S66-18602)

Chapter 3

Fig 3A	Apollo Launch Stack (NASA drawing)
Fig 3B	*Apollo 8* Earthrise (NASA 68-HC-870)
Fig 3C	Apollo CM Landing Config graphic (NASA Pub MSC-01856)
Fig 3D	Worldwide Abort Line chart (TF-140 Press Kit)
Fig 3E	USNS & ARIA (USAF via Chris Miller)
Fig 3F	*Apollo 8* Egress Training (NASA S68-53217)
Fig 3G	USNS *Vanguard* (USN via Carl Friberg)
Fig 3H	USNS *Mercury* (USN via Dwayne Day)
Fig 3I	ALOTS Variant (USAF via Robert Burns)
Fig 3J	*Apollo 11* ALOTS (NASA KSC-69PC-0413)

Chapter 4
- Fig 4A *Apollo 10* SIMEX helicopter (USN by Milt Putnam)
- Fig 4B *Apollo 8* Earthrise (NASA 68-HC-870)
- Fig 4C *Apollo 8* ALOTS (NASA S69-15592 via Charles Hinton)
- Fig 4D *Apollo 8* UDT CM Rope (NASA 68-HC-893)
- Fig 4E *Apollo 10* Parachute (USN by Milt Putnam)
- Fig 4F *Apollo 10* Recovery (NASA S69-21036 by Milt Putnam)
- Fig 4G *Apollo 10* Crew (NASA S69-20549)

Chapter 5
- Fig 5A MQF Lounge (by Stan Taylor, via John Blossom)
- Fig 5B Apollo Quarantine graphic (NASA image)
- Fig 5C Melpar MQF Cutaway drawing (Melpar Press Release 1969)
- Fig 5D MQF Bunk (by Stan Taylor, via John Blossom)
- Fig 5E MQF Control Panel (by Stan Taylor, via John Blossom)
- Fig 5F Melpar MQF-CM Connection drawing (Melpar Press Release 1969)
- Fig 5G BIG Suit (NASA S69-34538)
- Fig 5H MQF Aircraft Test (by Stan Taylor, via John Blossom)
- Fig 5I Water Egress Training (NASA S69-34885)
- Fig 5J JB & Rod Bass (by Stan Taylor, via John Blossom)
- Fig 5K MQF Build (by Stan Taylor, via John Blossom)
- Fig 5L MQF *Guadalcanal* HB (by Stan Taylor, via John Blossom)
- Fig 5M MQF *Wood* Fantail (by Stan Taylor, via John Blossom)
- Fig 5N MQF Aircraft Hold (by Stan Taylor, via John Blossom)

Chapter 6
- Fig 6 USS *Hornet* (USN via Pete Clayton)
- Fig 6B USS *Hornet* (USN by Robert Lawson)
- Fig 6C USS *Goldsborough* (USN by R.K. Hartkopp)
- Fig 6D USS *Hassayampa* (USN via Robert Carlisle)
- Fig 6E USS *Arlington* (USN via Jim Arnold)

Chapter 7
- Fig 7A C-1A Trader takeoff (USN by Robert Lawson)
- Fig 7B NT66 in Flight (USN by Milt Putnam)
- Fig 7C NT66 on Flight Deck (USN by Robert Lawson)
- Fig 7D NT66 in Hawaii (USN by Robert Lawson)
- Fig 7E E-1B Tracer (USN by Robert Lawson)
- Fig 7F VR-30 COD (USN by Robert Lawson)
- Fig 7G UDT San Diego Bay Training (USN via Joe Martinez)
- Fig 7H Clancy in Gulf (NASA S69-34881)
- Fig 7I ARRS HC-130H (USAF via USAF Museum)
- Fig 7J ALOTS NAV chart (USAF by Charles Hinton)
- Fig 7K ARIA EC-135N in flight (USAF via Chris Miller)
- Fig 7L Helicopter Navigational Aids (NASA S69-20592 by Milt Putnam)

Chapter 8
- Fig 8A *Hornet* Helicopter SIMEX (USN by Milt Putnam)
- Fig 8B *Hornet* Flight Deck Layout graphic (Carol Lee)
- Fig 8C GE Satellite Antenna (USN via Mike Wheat)
- Fig 8D ABC Van Onload (USN via Mike Wheat)
- Fig 8E Boilerplate Onload (USN via Mike Wheat)
- Fig 8F Helicopter UDT CM SIMEX (USN by Milt Putnam)
- Fig 8G NASA MQF Barge (USN via Tim Wilson)
- Fig 8H UDT *Hornet* SIMEX (USN by Michael Mallory)
- Fig 8I TV Camera on Tug (USN via Mike Wheat)
- Fig 8J King Neptune (USN via Tim Wilson)

Chapter 9
- Fig 9A Nixon arrival on *Arlington* (USN via Jim Arnold)

Fig 9B HMX-1 Helo C-133 (USMC via Dan Mcdyre)
Fig 9C Nixon JI Arrival (by Paul Hutchinson)
Fig 9D McCain COD JI (by Paul Hutchinson)
Fig 9E Nixon & USS *Carpenter* (USMC via Dan Mcdyre)
Fig 9F Nixon *Arlington* (USN via Jim Arnold)
Fig 9G Nixon Inspects Sailors (USN via Jim Arnold)

Chapter 10

Fig 10A *Apollo 11* Launch (NASA GPN-2000-000627)
Fig 10B *Apollo 11* Mission insignia (NASA S69-34875)
Fig 10C Pacific TLI Forces map (Carol Lee)
Fig 10D *Apollo 11* Re-entry Footprint chart (NASA Press Kit)
Fig 10E SIMEX BP Tilley (USN via Tim Wilson)
Fig 10F *Hassayampa* UNREP (USN via Chris Lamb)
Fig 10G *Hornet* Steaming (USN by Robert Lawson)
Fig 10H Air Ops Plan chart (USN *Hornet Cruise Report*)

Chapter 11

Fig 11A UDT CM retrieval (USN via Tim Wilson)
Fig 11B Nixon Exits *Marine One* (USN by Robert Lawson)
Fig 11C Nixon & Stonesifer MQF (USN via Mike Wheat)
Fig 11D Nixon & Group on Bridge (USN via Tim Wilson)
Fig 11E NASA CM Reentry graphic (NASA S68-41156)
Fig 11F NASA CM Chute Deploy graphic (NASA S66-10994)
Fig 11G CM in Stable 2 (USN by Milt Putnam)
Fig 11H Hatleberg Exits NT66 (USN by Milt Putnam)
Fig 11I Astronaut Decon in Raft (USN by Milt Putnam)
Fig 11J NT66 Astronaut Hoist (USN by Milt Putnam)
Fig 11K Overhead View of CM photo (USN via Tim Wilson)
Fig 11L Nixon & Paine on Bridge (USN via Tim Wilson)
Fig 11M NT66 Lands on Deck (USN via Dan McDyre)
Fig 11N Astronauts Exit NT66 (USN via Tim Wilson)
Fig 11O Astronauts Enter MQF (USN by Robert Lawson)
Fig 11P Hangar Bay View (USN via Mike Wheat)
Fig 11Q Nixon MQF Speech (USN by Robert Lawson)
Fig 11R Nixon Borman Joke (USN by Milt Putnam)
Fig 11S Nixon Congratulates Seiberlich (USN via Tim Wilson)
Fig 11T Clancy Removes BIG (USN by Michael Mallory)
Fig 11U Hornet Approaches CM (USN via Chris Lamb)
Fig 11V UDT Flower Appliqué (USN via Robert Lawson)
Fig 11W Crane Hoist CM from Helo (USN by Milt Putnam)
Fig 11X B&A Crane CM Hoist (NASA S69-21294)
Fig 11Y Chaplain Piirto (USN via Mike Wheat)

Chapter 12

Fig 12A *Apollo 11* Welcoming photo (USN by Robert Lawson)
Fig 12B NT66 Decon (USN via Tim Wilson)
Fig 12C STAR Balloon (USN via Tim Wilson)
Fig 12D STAR C-130 (USN via Tim Wilson)
Fig 12E NASA Moon Rock Package (USN via Tim Wilson)
Fig 12F C-1A Trader Launch (USN by Robert Lawson)
Fig 12G Astronauts in MQF (NASA KPC-69PC-484)
Fig 12H Hangar Bay Reenlistment (USN by Robert Lawson)
Fig 12I Letter Postmark Normal (author)
Fig 12J Letter Postmark Captain's (author)
Fig 12K Pearl Harbor Aerial View (USN via David Rush)
Fig 12L Pearl Harbor Pier Bravo (USN by Robert Lawson)
Fig 12M MQF Off-load (NASA S69-21881)
Fig 12N CM Hangar (USN by Jack LaBounty)

Epilogue

Fig 13A *Apollo 12* Mission insignia (NASA S69- S69-52336)
Fig 13B Gordon in NT66 (USN via Robert Lawson)
Fig 13C Davis MQF Welcome (NASA S69-22876)

Fig 13D   USS Hornet Museum (author)

Appendix A
Fig 14A   Armstrong Portrait (NASA MSFC-9018112)
Fig 14B   Armstrong Parade (USMC via Dan Mcdyre)
Fig 14C   Stonesifer & Aut (USN via Warren Aut)
Fig 14D   Stonesifer & Nixon (USN via Mike Wheat)
Fig 14F   Smiley in Flight Gear (USN via Chuck Smiley)
Fig 14G   Smiley & Seiberlich (USN via Chuck Smiley)
Fig 14H   Hatleberg in Flight (USN by Milt Putnam)
Fig 14I   Hatleberg in Gulf (NASA S69-34885)
Fig 14J   Blair Reporting (courtesy of Don Blair)
Fig 14K   Blair & Smith (courtesy of Don Blair)
Fig 14L   Putnam & NT66 (USN via Milt Putnam)
Fig 14M   Putnam & Nixon (USN via Mike Wheat)
Fig 14N   Seiberlich Portrait (USN via Heidi Seiberlich)
Fig 14O   Seiberlich & Nixon (USN via Tim Wilson)
Fig 14P   Seiberlich & Pirate (USN via Tim Wilson)

Appendix D
Fig 18A   Photographers in HB photo (USN via Robert Lawson)

Glossary of Acronyms & Abbreviations

The Navy, Air Force and NASA are massive bureaucracies that use very sophisticated technologies in the performance of their operations. Thus, the collision of these organizations during the Mercury, Gemini, and Apollo missions created a multitude of acronyms and abbreviations, not to mention mismatched definitions and different time reporting systems.

AFB	Air Force Base	CM	Apollo Command Module
AFETR	Air Force Eastern Test Range	CMP	Command Module Pilot
AFWTR	Air Force Western Test Range	COD	Carrier Onboard Delivery
AGMR	Major Communications Relay ship	CSM	Apollo Command Service Module
AIS	Apollo Instrumentation Ship	CTF	Commander, Task Force
ALOTS	Airborne Lightweight Optical Tracking System	CVS	Anti-Submarine Warfare Aircraft Carrier
AO	Fleet Oiler		
AOL	Atlantic Ocean Line	DD	Destroyer
ARIA	Apollo Range Instrumentation Aircraft	DDG	Guided Missile Destroyer
ARRS	Aerospace Rescue and Recovery Service	DDMS	DoD Manned Space Flight Support Office
ASW	Anti-Submarine Warfare	DoD	Department Of Defense
ATS	Applied Technological System		
		ELS	Apollo Earth Landing System
B&A	Boat and Aircraft	EOM	End of Mission
BIG	Biological Isolation Garment	EPL	Eastern Pacific Line
CIC	Combat Information center	GET	Ground Elapsed Time
CINCPAC	Commander in Chief, Pacific	GMT	Greenwich Mean Time
CLA	Contingency Landing Area		

HF	High Frequency radio	PLA	Primary Landing Area
HS	Helicopter Squadron	PRS	Primary Recovery Ship
HVT	USNS Huntsville		
		RED	USNS Redstone
IOL	Indian Ocean Line		
		S-IC	Saturn V first stage
KSC	Kennedy Space Center	S-II	Saturn V second stage
		S-IVB	Saturn V third stage
LM	Apollo Lunar Module	SARAH	Search And Rescue And Homing
LMP	Lunar Module Pilot	SATCOM	Satellite Communications
LORAN	LOng RAnge aid to Navigation	SIMEX	Simulation Exercise
LPH	Helicopter Carrier	SLA	Secondary Landing Area
LRL	Lunar Receiving Laboratory	SM	Apollo Service Module
		SRS	Secondary Recovery Ship
MAC	Military Airlift Command	SSB	Single Side Band radio
MCAS	Marine Corps Air Station	STAR	Surface To Air Retrieval
MCC	Mission Control Center		
MER	USNS Mercury	TACSAT	Tactical Communications Satellite
MPL	Mid-Pacific Line	T-AGM	Apollo Instrumentation class of ship
MQF	Mobile Quarantine Facility	TF-130	Task Force 130
MSC	Manned Spacecraft Center	TF-140	Task Force 140
MSFN	Manned Space Flight Network	TLI	Trans-Lunar Injection
MSTS	Military Sea Transport Service		
		UDT	Underwater Demolition Team
NAB	Naval Amphibious Base	UHF	Ultra High Frequency radio
NAS	Naval Air Station	USAF	United States Air Force
NASA	National Aeronautics and Space Administration	USMC	United States Marine Corps
NELC	Naval Electronic Laboratory Center	USN	United States Navy
NM	Nautical Miles	USNS	United States Naval Ship
		USS	United States Ship
PJ	Pararescue Jumper		

VAN	USNS Vanguard
VAW	Carrier Airborne early Warning squadron
VHF	Very High Frequency radio
WPL	Western Pacific Line
XRAY	Local ship time
ZULU	Greenwich Mean Time

Bibliography

At the end of a major sea cruise, most naval ships create a cruise book as a memento of that particular period. During the cruise, the ship's photographers and crewmen with cameras take candid photographs of officers and sailors doing their jobs as well as formal portraits of entire organizations.

The *Apollo 11* recovery was a very special, albeit short, cruise for the ship's crew with the at-sea period lasting just a little over four weeks. The *U.S.S. Hornet Apollo 11 Recovery Mission* book has become a major historical reference, the best available written record (with photographs) of the events that occurred during July 1969. It is also arguably the most unusual cruise book ever published by a naval vessel. Roughly twenty percent of the photographs were supplied by NASA and North American Rockwell and show men walking on the Moon and Apollo spacecraft in various phases of its flight.

The editor of the Cruise Book was Journalist Second Class Michael Wheat. The copy and managing editor was Ensign William Whitman. It was published by the Taylor Publishing Company in late 1969.

Other Books

Splashdown: NASA and the Navy by Don Blair, Turner Publishing Company, Paducah, Kentucky, 2004.

USS Hornet: A Pictorial History by Chuck Self, Turner Publishing Company, Paducah, Kentucky, 1997.

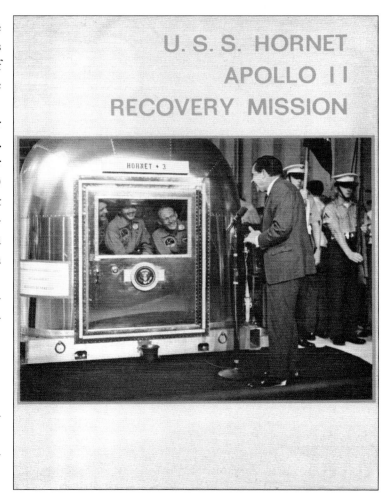

Other Sources

DoD: *Apollo 11* Press Kit – DDMS office, 1969.
NASA: *Apollo 11* Mission Report - MSC-00171, 1969.
NASA: *Apollo 11* Press Kit – 69-83K, 1969.
NASA: Apollo Recovery Ops Procedures - MSC-01856, 1969.
U.S. Navy: *Apollo 11* Press Kit - MSRF Atlantic (TF-140), 1969.
U.S. Navy: *Apollo 11* Ops Order 334-69 - MSRF Pacific (TF-130), 1969.
USS *Hornet*: Official *Apollo 11* Cruise Report, 1969.
USS *Hornet*: Official *Apollo 12* Cruise Report, 1969.

About the Author
Bob Fish

Bob Fish graduated from high school in Orlando, Florida in 1966 and attended the University of Virginia on a Naval ROTC scholarship.

After three years of college, he enlisted in the Marine Corps. He served most of his enlistment in Okinawa as a computer operator and shift supervisor at the U.S.M.C. Western Pacific data center. The center's primary responsibility was processing the supply requirements for the war effort in Vietnam. After being honorably discharged in 1971, he remained in the information technology field, holding increasingly important positions of responsibility at Blue Cross, Safeway Stores and Wells Fargo Bank. Of note, in 1978, he and a climbing partner made the first ascent of the Matterhorn in Switzerland that year. In 1980, he became the VP of Strategic Network Planning for the Bank of America and was the program manager for development of their worldwide communications network.

In 1984, he joined his first high tech Silicon Valley startup company and enjoyed the fast-paced environment. He spent the next twenty years working at Network Equipment Technologies (NET), Trident Data Systems, WheelGroup Corporation, Cisco Systems, Cloudshield Technologies and TKCIS. He has been involved in the founding or restart of five companies and three major divisions within existing companies, including being a founding investor of Pete's Brewing Company, the maker of the "Wicked Ale" that won national acclaim.

While working at NET, Bob was the program manager for the modernization of the White House communications system (during the Reagan era). After completing that task, he worked on a number of advanced government programs that required secure, high-speed networks. One of his companies, WheelGroup, created the first commercial network-based intrusion detection system and was acquired by Cisco in 1998.

In 2000, Bob decided to use his management and technical skills for community service and joined the Board of Trustees of the USS Hornet Museum in Alameda, California. He researched the Navy's role in recovering early NASA manned space-flights and created an award winning *Apollo 11* and *12* spacecraft recovery exhibit onboard the museum.

In 2004, he joined the Board of Directors of the Associated Airtanker Pilots (AAP) organization, which is comprised of aerial firefighting pilots nationwide, to help the families of fallen pilots obtain federal death benefits.

Bob and his wife Jennifer live in northern California. His personal interests focus on adventure travel and historical research. He has one son, Randy, who is currently active in the Peace Corps.

Index

A

Aerospace Rescue and Recovery Service (ARRS) 9, 16–17, 20–21, 26, 28–29, 40–41, 44, 56, 75–77, 79, 99, 105, 108, 113
Agnew, Vice President Spiro T. 90–91, 170
Airborne Early Warning Squadron VAW-111 70–71, 84, 107, 179
Airborne Lightweight Optical Tracking System (ALOTS) 28–29, 33, 39–40, 44, 77–78, 108, 111, 113
Air Force Eastern Test Range (AFETR) 8–9, 35, 75, 77–78
Air Force One 91–94, 126
Air Force Western Test Range (AFWTR) 9
Airstream Corporation 47, 50, 54–55
Alameda, Naval Air Station 72, 149
Aldrin, Edwin E. (Buzz) xvi–xvii, 5, 12, 56, 98–99, 102, 116–133, 134, 138–139, 149, 161, 163, 189–193
  seamount named after 104, 171
American Broadcasting Company (ABC) 83–85, 130, 141, 162–163, 179
American Standard. *See* Melpar
Anders, Bill 12, 37
Andersen AFB 77
Andrews AFB 91–92
Anti-Submarine Warfare 59, 68–71, 157, 168–169
  Project UPTIDE 60–61, 146, 149, 169
Apollo 1 6, 36, 178
Apollo 7 12, 37, 179
Apollo 8 xviii, 12, 30, 37–40, 42–44, 66, 68–70, 94, 125, 153, 155–157, 165–166, 179
Apollo 9 12, 14, 30, 40, 157
Apollo 10 12, 36, 41–44, 53, 66, 69–70, 72, 75, 79, 83–84, 90, 99, 121, 156–159, 166, 179
  Command Module *Charlie Brown* 41–43, 158
Apollo 11 xii, xvii–xviii, 5, 8, 10, 12, 33–34, 51, 53, 56, 59–61, 64, 66, 67, 69–72, 77–78, 81–89, 91–95, 97–107, 147, 149, 151–155, 159–161, 163–174, 179–181, 189–193, 196–198
  Command Module *Columbia* 78, 82, 99, 101–102, 108, 111–119, 127–131, 134, 136–137, 141, 143–145, 151, 160–161, 164, 170, 173, 180–181, 197
    kapton foil 129, 131, 141
  Lunar Module Eagle 102
  postal cachet 139–140
  seamount named after 104, 171
Apollo 12 xi–xii, xvii, 12, 51, 56, 69–70, 72, 146–148, 153, 155, 169, 172, 174
  Command Module Yankee Clipper 147, 169
Apollo 13 12, 57, 69–70, 147–149, 156
Apollo 14 12, 50, 51, 56–57
Apollo 15 12
Apollo 16 12, 72
Apollo 17 12
Apollo Flights, Unmanned 36–37, 153
  Apollo 4 36
  Apollo 5 37
  Apollo 6 37
  Apollo-Saturn 201 (AS-201) 36
  Apollo-Saturn 202 (AS-202) 10–11, 36, 60
  Apollo-Saturn 203 (AS-203) 36
Apollo Instrumentation Ships (AIS) 27–28, 31–35, 99–100
Apollo missions, generally xv–xviii, 12, 15, 22–35, 36–44, 67–80
Command and Service Module (CSM) 23–25, 36–37, 79–80
  boilerplate 25, 28, 53, 56, 81, 83–84, 87, 102, 160, 163, 179
  Earth Landing System (ELS) 23–25, 37, 43, 80, 157
  Lunar Module 23, 37, 40–41, 45
Apollo Range Instrumentation Aircraft (ARIA) 27–29, 33–35, 40, 44, 75, 77–78, 108, 111, 113, 137
Apollo - Soyuz 12
Arlington (AGMR-2) 40, 59, 64–66, 90, 92, 94–95, 103–104, 106, 109, 114, 126
Arlington Cemetery 144
Armstrong, Neil xiii, xvi, 5, 12, 20–21, 53, 56, 98–99, 102, 116–133, 134, 138–139, 141, 151–152, 161, 163, 167, 189–193
  seamount named after 104, 171
Atlantic Ocean Line (AOL) 25–26, 28
Aurora 7 12, 16–17, 75, 155, 178
Aut, Warren 147, 153

B

Barbers Point, Naval Air Station 87
Baron, Oakley 77
Barrett, Richard 194
Bass, Rod 54
Bean, Alan xi–xii, 12, 56, 146–148
Bendix Corporation 33–35
Benton, Gregory 195
Biological Isolation Garments (BIG suit) 46, 50–53, 74, 107, 110, 116–117, 121–123, 127, 129, 147, 159–161, 170

Blair, Don xviii, 85, 89, 162–164
Blossom, John 54–57
Boisvert, Louis 42–43
Borman, Frank 12, 37–39, 93–94, 125, 192–193
Brand, Vance 12
Bucklew, Mitchell 195
Bulkeley, J.D. 146
Burns, Governor John 141

C

C-1A Trader 40, 67, 71–72, 84, 92–93, 103–104, 127, 134, 137, 179
C-141 Starlifter 52, 56–57, 137, 141, 154, 181
Cape Canaveral xv, 6, 9, 15. *See also* Kennedy Space Center
Carpenter (DD-825) 59, 93–94, 104
Carpenter, Scott 12, 16–17, 155, 178
Carpentier, Dr. William 102, 109, 118–119, 121–123, 136, 139, 152, 194
Carr, Gerald 12
Carrier-Onboard-Delivery (COD) 71–72, 84, 92–93, 94, 103–104, 141, 179
Ceausescu, Nicolae 90
Cernan, Eugene 12, 41, 44, 158
Chaffee, Roger 36
Chesser, Wesley 42–43, 116, 158–160, 194
Coggin, Bob 39
Collins, Michael 12, 51, 56, 98–99, 102, 116–133, 134, 138–139, 161, 163, 191–193
  seamount named after 104, 171
Columbia Broadcast System (CBS) 85, 162–163
COMNAVAIRPAC band 88, 121–125, 138–139, 141
Conn, George 194
Conrad, Charles (Pete) xi–xii, 12, 20, 56, 146–148
Cooper, Gordon 12, 17–18, 20, 178
Coronado, Naval Amphibious Base 14, 72, 160
Cox, George 15–16
Cruse, Carl 44
Cuban Missile Crisis xv, 5, 178
Cunningham, Walt 12, 37

D

Dalpiaz, Paul 195
Davis, RADM Donald C. xi, 10, 109, 134, 147–148
Defense Meteorological Satellite Program (DMSP) 106
DoD Manager for Manned Space Flight Support Office (DDMS) 7–8
DoD, manned space flight support xiii, xvi–xviii, 5–10, 13–21, 153
  Apollo Missions 23–30, 36–44
Dominguez, Abram 195
Duke, Charles 12
Duncan, Larry 195

E

E1-B Tracer 70–71, 84, 103–104, 108, 134, 179
East Pacific Line 25–26
EC-135N ARIA 27, 33, 35, 108
Eisele, Donn 12, 37
Ellington AFB 56, 137, 141, 154, 181
Essex (CVS-9) 12, 37, 151
Evans, Ronald 12

F

Faith 7 12, 14–15, 17–18, 178
Fasi, Mayor Frank 141
Fifield, John 39
Flanagan, Dick 39
Fleet Composite Squadron VC-1 87
Fleet Tactical Support Squadron VR-30 40, 71–72, 84, 126, 179
Free, Charles 195
Freedom 7 12, 13–15, 162, 177
Friendship 7 12, 178

G

Garriott, Owen 12
Gemini 3 12, 61, 178
Gemini 4 12, 19–20, 61, 178
Gemini 5 12, 20, 61

Gemini 6A 12, 178
Gemini 7 12, 178
Gemini 8 12, 20–21, 64, 75, 141, 178
Gemini 9 12, 64, 163
Gemini 10 12, 163, 178
Gemini 11 xi, 12, 163, 178
Gemini 12 12, 178
Gemini missions, generally xv, 10, 12, 19–21, 23, 26–27, 74, 153
General Electric (GE) 82–85, 134, 179
Gibson, Robert 12
Gilbert Islands 108, 113
  Abemama Atoll 111
Glenn, John 12, 178, 187
Goldsborough (DDG-20) 59, 61–64, 87, 100–101
Gordon, Richard xi–xii, 12, 56, 146–148
Grissom, Gus 12, 15–16, 36, 177
Grumman Aircraft 71
Guadalcanal (LPH-7) 12, 40, 55, 157
Guam 66, 77, 91, 104, 126
Guam (LPH-9) xi, 12

H

H-34 Seahorse 14–16, 68
Haise, Fred 12
Haldeman, Robert 93
Hassayampa (AO-145) 59, 63–64, 103, 165
Hatleberg, Clancy 51, 53, 74–75, 109, 114–119, 127, 129, 151–152, 159–161, 170, 194
HC-130H Hercules 26, 40–41, 56, 75–77, 79, 134–135
  Surface-to-Air-Recovery (STAR) 40, 134–135, 181
Helicopter #66 (*Recovery One*) 39, 42–44, 68–70, 79, 104, 105, 107, 109, 114–115, 119–123, 130, 134–135, 147, 151, 153, 157–158, 160–161, 165–166, 181, 194
Helicopter ASW Squadron 3 157
Helicopter ASW Squadron 4 xviii, 36, 39, 42–44, 67–70, 81, 84–85, 87, 109, 119, 121, 147, 151, 153, 156–158, 165–167, 179–180
Helicopter ASW Squadron 5 15

Hickam AFB 77, 137, 141–142, 147, 154, 181–182
Hill, Curtis 194
Hinton, Charles 77, 113
Hirasaki, John 102, 121, 134, 136, 194
Hoback, Kenny 128, 194
Holt, Joe 140
Hornet Apollo 11 Recovery Mission 105, 198, 205
Hornet (CVS-12) xi–xiii, xvii–xviii, 10–12, 36, 56, 58–61, 64, 66, 67–68, 71–72, 75, 77, 81–89, 90–92, 97, 99–107, 108–133, 134–144, 146–149, 151–152, 154, 157, 160–161, 163–166, 168–174, 179–181, 190–193, 196–198
Huntsville (T-AGM 7) 99–100
Huyett, Larry 20–21

I

Indian Ocean Line 25–26
INTELSAT IV 86
Interagency Committee for Back Contamination (ICBC) 45, 154–155
Intrepid (CVS-11) 12, 16–17, 61
Irwin, James 12
ITT 33–35
Iwo Jima (LPH-2) 12

J

Jahncke, Ernie 147
James, Ben 101
Jernigan, Clarence 39
John R. Pierce (DD-753) 12, 17
Johnson, Bruce 194
Johnson, James 195
Johnson, President Lyndon B. 39, 178
Johnson Space Center. *See* NASA, Johnson Space Center
Johnston Island 59, 61, 66, 91–94, 97, 104, 106, 114, 126–127, 137, 180–181
Jones, Don xviii, 39, 69, 118–119, 156–157, 170, 194
Jones, Lee 109, 165, 195, 196–198
Jones, MGEN David M. 8

K

Kearsarge (CVS-33) 12, 17–18
Kennedy, President John F. xii–xiii, 3, 6, 15, 111, 144, 172, 174, 177–178, 183–189
Kennedy Space Center 36–37, 92, 98–99. *See also* Cape Canaveral
Kerwin, Joseph 12
Kissinger, Henry 93–94, 164
Klepak, George 195
Knapp, Dick 194
Koons, Wayne 15
Korean War 151

L

Lake Champlain (CVS-39) 12, 15, 20
Lamb, Chris 81, 106, 141, 144, 194
landings, passive water-based 6
 versus ground-landing xi, 6
Larson, Charles 87, 91–92, 126–127
Lautermilch, Paul 61
Lawson, Robert 198
Leonard F. Mason (DD-852) 12, 20–21
Lewis, Jim 15–16
Liberty Bell 7 12, 15–16, 177
Little Creek, Naval Amphibious Base 14
Lousma, Jack 12
Lovell, Jim 12, 29, 37
lunar pathogens. *See* Moon germs

M

Mallory, Michael 42, 116, 159–160, 194
Marine Air Group 26 (MAG-26) 15
Marine One 82, 90–91, 94, 97, 104, 108–109, 126
Marines, U.S.
 manned space flight program support 13–21
Mattingly II, Thomas Kenneth 12
Mayo (DD-422) 168
McBee, Keith 85, 141, 163
McCain, Admiral John S. xii, 92–93, 94, 109–111, 126–127, 141, 147, 154, 169, 180–181
McCain, John (Senator) 127
McCoy AFB xv
McDivitt, Jim 12, 20, 40
McDonnell-Douglas 35
McLaughlin, John 195
McManus, RADM Philip S. 10
McNally, John 194
McPhee, C.E. 92
Melpar 47, 49–50, 54
Merchant Marine Academy, U.S. xvii, 60, 168–169
Mercury missions, generally xv, 9, 10, 12, 13–18, 23, 27, 153
Mercury (T-AGM 21) 31–35, 99–101
Mid-Pacific Line (MPL) 25–26, 28, 61, 72, 87, 101, 180
Midway Island 66, 92, 172
Military Sea Transportation Service (MSTS) 33–35
Miller, Michael 195
Mission Control Center. *See* NASA, Mission Control Center
Missouri (BB-63) 149, 168
Mitchell, Edgar 12, 57
Mobile Quarantine Facility (MQF) xii, 30, 45–57, 84, 86, 102, 110, 121–125, 129, 131–132, 134, 136–139, 141, 144, 147–148, 154, 164, 173, 180–181
 microwave oven 48, 56, 138
 transfer lock 47–48, 55–56, 136, 138
Moody, Joe 109
Moon germs 19, 45–57, 74–75, 81, 88, 106–107, 109, 116–127, 134–135, 147, 149, 152, 154, 164, 168, 170–171. *See also* Biological Isolation Garments (BIG suit); *See also* Mobile Quarantine Facility (MQF)
 biological decontamination process 51–53, 116–119, 121–122, 159–161, 170, 181
Moon rocks 45–47, 49, 56, 71, 81, 127, 134, 136, 136–137, 169, 171, 173, 181
 lunar sample return containers 136
Mooney, Howard 106, 109
Moore, Glenn 20–21
Muehlenbach, Terry 195

Murphree, Hugh 66, 94
Mutual Broadcasting System 82, 84–85, 89, 162–164, 179

N

N1 Rocket, Soviet 6
NASA xvi, 179
    Capsule Communications (CAPCOM) 32–33, 106
    Deep Space Network (DSN) 27, 31
    Johnson Space Center (JSC) 7, 16, 52–53, 56, 104, 144, 154, 181, 196
        Lunar Receiving Laboratory (LRL) 45, 49–50, 136–137, 141, 144, 154, 181
    Langley Research Center 13
    Manned Space Flight Network (MSFN) 8, 31–35
    Mission Control Center xii, 8, 27–29, 35, 86, 106, 153
    relationship with DoD xiii, xvi–xviii, 5–10, 13–21, 23–30
National Broadcasting Company (NBC) 85, 162–163
Navy Navigation Satellite System (TRANSIT) 41, 84–85, 106, 146, 171, 178
Navy, U.S.
    Task Force 130 xi, 10, 28, 64, 66, 70, 75, 77, 84, 86, 92, 101, 107, 108–109, 147–148, 165, 179
    Task Force 140 9–10, 28, 100
    Underwater Demolition Teams (UDT), generally xi, 13–21, 28–30, 67, 69, 80, 81, 83, 85, 106–107, 141, 147, 158, 159, 163, 198. *See also* Individual UDT Units, e.g., UDT-11, etc.
    decontamination swimmer 50–53, 74, 107, 160
Neal, Eldridge 20–21, 141
Neptune, King 89, 172
Nessen, Ron 85, 163
New (DD-818) 59
New Jersey (BB-62) 149
New Orleans (LPH-11) 12, 57
New River, Marine Corps Air Station 13

Nixon, President Richard M. 5, 59, 66, 81–82, 87, 90–95, 97, 102–104, 107, 109–126, 133, 152, 154, 164, 166, 168–171, 173–174, 179–181, 189–193
    Nixon-Thieu summit 66, 92
Noa (DD-841) 12

O

Okinawa xvi, 20–21, 60, 207
Okinawa (LPH-3) 12
Ozark (MCS-2) 59

P

Pago Pago 147
Paine, Thomas 93, 109, 120, 164
Pararescue Jumpers (PJs) 9, 12, 20–21, 26, 28, 56, 75–77, 141
    41st Air Rescue Squadron 16–17
Patch, George 108, 113, 194
Patrick, Andy 195
Patrick AFB 35, 75
Pearl Harbor xii, 86–87, 91, 114, 129, 141–144, 147, 152, 155, 160, 170, 179–181, 197. *See also* Hickam AFB
Phoenix Islands 61, 89
Piirto, Chaplain John 102–103, 126, 132–133, 181, 193
Pogue, William 12
Powers, Dick 194
Princeton (LPH-5) 12, 36, 41–44, 53, 61, 66, 79, 83, 157, 158
Pueblo incident 5
Putnam, Milt 43, 109, 165–167, 195, 196–198

Q

Quemoy-Matsu 64

R

Rambaut, Dr. Paul 138

Randolph (CVS-15) 12, 15–16
Rankin, George 157
RCA 33–35
Redstone (T-AGM 20) 31–35, 99–100
Reinhard, John 16
Retriever, M/V 51, 53
Rice University 3, 185–189
Richmond, Donald 195
Robnett, Stanley 119, 194
Rogers, William 93–94, 193
Rohrbach, Bob 195
Roosa, Stuart 12, 56
Ryan, Bill 163

S

Sabye, Rolf 106
Safire, William 190
Saipan (CVL-48). *See* USS Arlington
Salamonie (AO-26) 168
Salinan (ATF-161) 59
Sample, Edward 109
Saturn V xvi, 5–6, 6, 22, 36, 40, 92, 98–99
Schirra, Wally 12, 37, 178
Schmitt, Harrison 12
Schwab, Don 39
Schweickart, Rusty 12, 40
Scott, David 12, 20–21, 40, 141
Search and Rescue (SAR) 19, 30, 60, 68–69, 80, 134
Search and Rescue and Homing (SARAH) 79–80
Secret Service 66, 67, 87, 91–92, 103, 109, 110, 126, 173
Seiberlich, Captain Carl J. xii–xiii, xvii–xviii, 60–61, 81, 85–87, 89, 90–91, 104, 109, 114, 118, 126, 128, 134, 138–141, 144, 147–149, 157, 164, 168–174, 179, 194
Seventh Fleet 64
SH-3 Sea King 17, 19, 42–43, 68–70, 79–80, 84, 103, 107, 116–118, 119–122, 155–158, 167, 179
Shepard, Alan 12, 13–16, 56, 162, 177
Shockley, Thomas 77
Sigma 7 12, 14–15, 178

Sikorsky Aircraft 70
Simulated Recovery Exercise (SIMEX) 80, 86–89, 102, 106–107, 147, 154, 159–160, 165, 170, 174, 179
Skylab 2 12
Skylab 3 12
Skylab 4 12
Skylab missions, generally 153
Slayton, Donald "Deke" 12
Slider, Glen 158
Smiley, Charles B. 42–44, 69, 79, 156–158
Smithsonian Institute 144
Solomon Islands 99–100
Stafford, Thomas 12, 41, 44, 158
Stonesifer, John 17, 37, 39, 101, 109–110, 121–122, 153–155, 170, 194
Strawn, William 195
Stullken, Don 16, 44, 121, 161, 170, 194
Swigert, John 12

T

Tactical Communications Satellite (TACSAT) 40, 82, 84, 86
Taylor, Stan 54
Thieu, President Nguyen Van 66, 92
Ticonderoga (CVS-14) 12, 169
Tolleson, Robert T. 10
Townsend, Dallas 85, 121, 163
Trans-Lunar Injection (TLI) 22, 25–28, 61, 77–78, 99–101, 147, 180

U

UDT-11 14–15, 42–44, 72–75, 85, 87, 108–109, 114–120, 128–133, 151–152, 159–161
UDT-12 14–15, 39–40, 72
UDT-13 14, 72–75, 147
UDT-21 14
UDT-22 14
Underwater Demolition Teams (UDT), generally. *See* Navy, U.S.
underway replenishment (UNREP) 64, 103

USS Hornet Museum xvii–xviii, 149–150, 169, 172
  Aircraft Carrier Hornet Foundation 149

V

Vanguard (T-AGM 19) 27, 31–35, 99–100
Via, Joseph 195
Vietnam War xii, xvi–xvii, 5–6, 14, 58, 60–61, 64–65, 68, 72, 81, 90, 127, 165–166, 168–169, 172–173, 177–179
Voice of America 84, 179

W

Waits, Jack 64
Wake Island 10–11, 36, 60, 77, 108
Walker, Scott 44, 158
Wasp (CVS-18) 12, 20, 163
Weitz, Paul 12
Western Pacific Line 25–26
Western Union International (WUI) 82–87, 134
Wheat, Michael 205
White, Edward 12, 20, 36, 178
White House Communications Agency (WHCA) 66, 67, 81, 87, 91, 103
White House Press Corps 90, 104
Whitman, William 205
Whitten, Charles 195
William M. Wood (DD-715) 56
Wilson, Tim 194
Wolfram, John 116, 129, 159–160, 194
Wood, Norvel 119, 194
Worden, Alfred 12
World War II xiii, xv, xvii, 27, 37, 58–60, 94, 149, 151, 155, 168, 172–173

Y

Yorktown (CVS-10) 12, 30, 37–40, 61, 66, 155–156, 165
Young, John 12, 41, 44, 158

Z

Ziegler, Ron 93